Columbia University

Contributions to Education

Teachers College Series

No. 718

AMS PRESS
NEW YORK

AMERICAN AGRICULTURAL PROBLEMS IN THE SOCIAL STUDIES

SOME IMPORTANT AGRICULTURAL PROBLEMS AND RELATED GENERALIZATIONS THAT SHOULD BE CONSIDERED IN THE GENERAL CURRICULUM OF URBAN AND RURAL SCHOOLS

By KENNETH E. OBERHOLTZER, Ph.D.

TEACHERS COLLEGE, COLUMBIA UNIVERSITY
CONTRIBUTIONS TO EDUCATION, NO. 718

Published with the Approval of
Professor Jesse H. Newlon, Sponsor

BUREAU OF PUBLICATIONS
TEACHERS COLLEGE, COLUMBIA UNIVERSITY
NEW YORK CITY
1937

Library of Congress Cataloging in Publication Data

Oberholtzer, Kenneth Edison, 1903-
American agricultural problems in the social studies.

Reprint of the 1937 ed., issued in series: Teachers College, Columbia University. Contributions to education, no. 718.
Originally presented as the author's thesis, Columbia.
Bibliography: p.
1. Agriculture--Study and teaching--United States.
2. Social sciences--Study and teaching--United States.
I. Title. II. Series: Columbia University. Teachers College. Contributions to education, no. 718.
S533.O3 1972 375'.3381'0973 73-177125
ISBN 0-404-55718-X

Reprinted by Special Arrangement with Teachers College Press, New York, New York

From the edition of 1937, New York
First AMS edition published in 1972
Manufactured in the United States

AMS PRESS, INC.
NEW YORK, N. Y. 10003

ACKNOWLEDGMENTS

Grateful acknowledgment is made to the sponsor of this dissertation, Professor Jesse H. Newlon, and to the advisers, Professors Herbert B. Bruner, Edmund deS. Brunner, and Harold F. Clark. They have been gracious critics and friendly counselors.

To Professor Carter Alexander I owe much for suggestions and encouragement from the inception to the close of this study. To my wife, Florence Oberholtzer, and to my father, E. E. Oberholtzer, I am indebted for continued faith and encouragement in the work.

I am also grateful to many other individuals who helped with their criticism of the manuscript. Librarians were kind in making source materials available, particularly at the Agricultural and Mechanical College of Texas, Teachers College, Columbia University, Texas Technological College, and the University of Chicago.

K. E. O.

CONTENTS

CONTENTS

CHAPTER I

THE PROBLEM FOR STUDY

THE American *agricultural industry** is confronted with many problems of such great importance that all citizens should have a more adequate understanding of them. All citizens will have the opportunity to acquire more adequate knowledge of them when they are considered in the public schools. This study purposes to define and state these problems in such manner as to be helpful in the construction of modern courses of study or curriculum units, for both urban and rural schools; and, further, to present, regarding these problems, a mass of generalizations from worthy sources. There is at present no single publication giving curriculum workers such materials in a form that can be readily used. Widespread activity in the construction of public school curricula has caused a demand for new printed materials of this kind. For supplying this demand there are here presented such materials, selected through research techniques of recognized scientific value.

The materials will probably be of most value to those curriculum workers who may wish to develop courses for our American secondary schools. How and where they may be used is treated more fully in Chapters IV and V; the procedure followed in finding and stating the problems is described in Chapter II; the problems together with related generalizations are presented in Chapter III.

This study is not concerned with materials for the preparation of young men and women in the vocation of farming or with the direct vocational training of any group of our American

* *Agricultural industry*. This term is used because agriculture is a distinct type of occupation that employs much capital and labor. It employs machines; its products are becoming standardized, it is truly a part of our great industrial scheme.

population. Rather, it is concerned with materials that may be of assistance in the preparation of young men and women for more effective living in our modern society, and for a clearer understanding of the forces that affect their living.

School people often divide the course of study into two parts: the general and the vocational. Vocational education is that which prepares students for successful participation in fields of service, for earning a livelihood through employment in an occupation. General education includes all other activities in the school, provided for all pupils in order to develop useful citizens. The materials offered in this study would no doubt be of some worth for vocational education, but they are designed mainly for general education.

A NEED FOR UNDERSTANDING CONTEMPORARY PROBLEMS

Greater social, economic, and political benefits will accrue to our American people when there is a clearer understanding and truer appreciation of occupational groups. For the various problems of one group are so enmeshed with those of other groups that it is very difficult to employ any line of demarcation of interests. This interdependence of vocational groups is one of the chief characteristics of our modern society, and the growth of our great urban population has tended to accentuate this characteristic in America. Our urban population is dependent upon the rural population for essential food products and for products which furnish raiment and shelter; our rural population is dependent on the urban population for manufactured articles, and for markets for agricultural surplus. This agricultural surplus (due to the efforts of producers) leads to a distributors' problem, and this in turn to a consumers' problem; all are involved. Each group has its problems that affect other groups, and each group needs to know something of the other's problems before much mutual benefit can be derived.

The agricultural group is one of our major population and occupation groups. Those engaged in the agricultural industry comprise about one-third of our population. Directly dependent upon the welfare of agriculture we find other large groups, such as manufacturers and distributors of farm implements, distributors of farm products, manufacturers and distributors of commercial fertilizer, rural bankers, and rural merchants. The interests of one-third of our population, our agricultural producers, and of those who depend on them for their material welfare, must be considered by both rural and urban people.

Periodically in the history of our nation there has been evidence of general unrest among rural people, with charges from rural leaders that agriculture has not been getting its just share in the material and cultural benefits of the day. Often writers have referred to this general situation as the farm problem. To imply that there is only one "farm problem" is to over-simplify what is very complex. *There is no one farm problem, but there are many problems relating to the agricultural industry, problems which are stated later in this study and which merit the consideration of pupils in our public schools.* When pupils recognize and study some of the problems of our agricultural group, they will recognize and study problems of one of our largest and most essential population and occupation groups.

THE PUBLIC SCHOOLS AND PROBLEMS OF SOCIETY

One of the primary aims of schools is the induction of boys and girls into the group life of our civilization. If our public school curriculum is to be true to this aim, then it must include a study of the major problems of that life.

The concepts and ideas necessary for an understanding of our society must be built gradually from childhood through youth and the years of adult life. Education must be regarded as a process of growth in social understanding. Only through the consideration of domestic and world problems can come the social and intelligent outlook commensurate with the creation of a social order free from the suffering and injustices characteristic of conditions today. Moral problems, aesthetic problems, the problem of the more equitable distribution of wealth, problems of international relationships, of peace and war, problems pertaining to the relation of classes and races—all of these vital problems of contemporary life—must become subjects of study in our schools.[1]

[1] *Report of the Survey of the Schools of Chicago, Illinois,* Vol. III, p. 16, 1932.

The school justifies itself in the measure in which it equips individuals to engage in the activities of life with effectiveness and satisfaction. Only as the knowledge, habits and attitudes, and appreciations developed in school are operative in meeting the problems of life are they of any worth.[2]

Much of what is now taught is too largely conventional and all too remote from life. The idea that education consists in the acquisition of stated subject matter must give way to the study of problems vital within the lives of the young people and to the undertaking of enterprises significant within the community. Only in such way can we hope to get the needed intelligent thinking about social affairs or build adequate social attitudes. As far as the age of the pupils will permit, they must become intelligently critical of our, of their, social life and institutions.[3]

And so the problems of American agriculture should become objects of study in our public school curricula, not only in rural schools but in urban schools as well.

AGRICULTURAL PROBLEMS FOR RURAL AND URBAN SCHOOLS

It is known that not all rural-born children will later live in rural areas, nor will all urban-born children later live in urban areas. There is a constant shifting of population between these areas; consequently, for (all) girls and boys living in rural areas education should not be for living in rural areas alone, nor should (all) girls and boys living in urban areas be educated for living in urban areas exclusively. If there is to be an adequate consideration of this interchange of population and the interdependence of our various population groups, it becomes apparent that there should be much educational material common for rural and urban schools. Rural schools will be considering problems of rural and urban life; urban schools will be considering problems of urban and rural life. If this is to be true, then both urban and rural schools should consider problems of American agriculture. This must be so if they are to consider some of the "vital problems of contemporary life," and if they are to be regarded as of assistance in the "process of growth in social understanding."

[2] F. G. Bonser, *The Elementary School Curriculum,* p. 76. 1920.
[3] W. H. Kilpatrick, *Education and the Social Crisis,* p. 80. 1932.

CHAPTER II

THE PROCEDURE USED IN SOLVING THE PROBLEM

WHAT type of printed source materials would be most useful to secondary school curriculum workers when considering the agricultural industry? An examination of other printed studies reveals that "problems" and "generalizations" or "concepts" are the most frequently used materials of assistance to workers.

The problem is well-nigh universal in every field of human endeavor, educational and vocational, for whenever the adjustment of thought to a fact or a fact to a thought is involved, there the problem lies close at hand.[1]

Problems, stated or unstated, furnish the basis for most of our reflective thinking in life situations outside the classroom and they have also been the cause of much of our thinking in classrooms.

Not the learning of texts but the solving of problems is what we need. Our materials must be organized around issues, problems, unanswered questions which the pupil recognizes as important and which he really strives to unravel.[2]

The most significant question which can be asked, accordingly, about any situation or experience proposed to induce learning is what quality of problem it involves.[3]

For these reasons problems were selected as one of the basic curriculum materials for this study. But problems alone, however well stated, would be of little value to curriculum workers. There must be some qualifying materials which will assist in arriving at solutions. There must be some store of meanings before a problem can be recognized as such and before solutions can be suggested. In order, therefore, to give some measure of control

[1] Charles De Garmo, *Principles of Secondary Education,* Vol. II, p. 24. 1908.

[2] Harold Rugg, *The Social Studies in the Elementary and Secondary School,* Part II, p. 20. 1926. [3] John Dewey, *Democracy and Education,* p. 182. 1916.

in the recognition and solution of the problems it was decided to state significant generalizations or meanings related to them.

> Increase of the store of meanings makes us conscious of new problems, while only through translation of the new perplexities into what is already familiar and plain do we understand or solve these problems. This is the constant spiral movement of knowledge.[4]

Consequently, both problems and generalizations were selected as the basic materials to be defined and stated for the benefit of curriculum workers.

HOW LOCATE NEEDED PROBLEMS AND GENERALIZATIONS?

In order to arrive at the needed problems and generalizations an analysis of certain published materials was decided on as the best means for this study. Great masses of statistics are available for the student to interpret and thereby locate problems and generalizations; one might interview large numbers of selected individuals; and there are other techniques which might be followed. The technique of analysis used in this study and the type of materials analyzed are similar in many respects to those of earlier studies made for the development of social science curriculum materials. This present study was attempted partly because the findings of these earlier studies were judged to be inadequate insofar as they considered the agricultural industry.[5]

The published materials finally selected for analysis represent the thought of men and women who are recognized as leaders in critical interpretation of the problems of our American agricultural industry, and who presented their findings to the American people in current magazines, books, and reports. An analysis of such published materials may be justified on at least two bases. The first has been expressed as follows:

[4] John Dewey, *How We Think*, p. 120. 1910.

[5] Neal Billings, *Determination of Generalizations Basic to the Social Studies Curriculum.* 1929.

John A. Hockett, *A Determination of the Major Social Problems of American Life.* 1927.

Baldwin Lee, *Issues in the Social Studies.* 1926.

Marian Olive Lerrigo, *Health Problem Sources.* 1926.

Hyman Meltzer, *Children's Social Concepts.* 1925.

This furnishes one valid criterion, surely, that people need to know what they read about, and, by implication, what they talk about.[6]

The second has been well stated by Rugg and Hockett:

Whereas the essential skills and much of the factual knowledge of the curriculum can be discovered by analysis of what people do, the more fundamental insights, problems, and principles will be found only by a study of expert thought and feeling . . . such focal matters as the fundamental problems of contemporary life, the chief characteristics of American culture, basic causal relationships are to be discovered by a study of what the frontier thinkers think people should be able to do.[7]

WHAT IS A PROBLEM?

Before an analysis could be made for problems and for generalizations, it was necessary to establish certain guiding definitions. Webster's Dictionary presents this definition of a problem: "A question proposed for solution; a matter stated for examination or proof; hence, a matter difficult of solution or settlement; a doubtful case or question." So in the analysis of materials for problems, questions stated were recorded as problems, such as: "What should be done?" "How could this be done?" "Why should this be done?" Whenever a direct statement of problem was included in the materials, it was recorded as a problem. For example, "The real problem lies in the number of agricultural workers which we have on farm lands." Naturally there were a large number of implied problems in every article or book read, but these were not recorded in the first analysis because only stated materials were sought. A record of all implied problems would introduce a large element of opinion into the analysis, thus reducing the element of objectivity. Actually, after the analysis for problems and generalizations was completed, there were groups of generalizations which raised questions that were not stated directly in the materials read. Where these generalizations did imply, collectively, a "doubtful case or question," a question was formu-

[6] Philip W. L. Cox, *Curriculum Adjustment in the Secondary School*, p. 81. 1925.

[7] Rugg, Harold and Hockett, John, *Objective Studies in Map Location*, pp. 49-50. 1925.

lated. This was then included as one of the problems. Mere description or narration of a situation or condition was not recorded as a problem, for a problem demands a solution.

WHAT IS A GENERALIZATION?

When one again refers to Webster's New International Dictionary for a definition, there "to generalize" is "to derive or induce (a general conception, or a general principle) from particulars . . . to consider with reference to general or generic relations." A generalization is the product of generalizing. As used in this study, generalizations include principles, laws, ideas, and general views which were derived from a consideration of many particular cases or instances. Generalizations, as here used, are not to be confused with such exact laws as are found in the realm of the physical sciences (such as the laws relating to falling bodies, in physics). They are more nearly significant ideas which students of our problems have arrived at. Such generalizations give meaning to problems stated or implied.

CRITERIA FOR THE SELECTION OF AGRICULTURAL PROBLEMS AND GENERALIZATIONS

The aforestated definitions of a problem and a generalization set a general pattern for analysis, but further delimiting characteristics of each were necessary before the actual analysis for them could be made. One of these characteristics has been mentioned in Chapter I, namely, that this study is concerned primarily with problems of interest to people in connection with their living rather than with their making a living. So not all problems and generalizations appearing in the chosen source materials would necessarily be recorded. In order to select the needed problems and generalizations, certain criteria were decided on as the bases. Each problem or generalization selected should satisfy the following criteria:

1. *It should be of national or international importance.* The problem or generalization should be one that would be considered by Congress, by the American Farm Bureau, or by some national planning body, for example.

2. *It should affect the welfare of our people as a group.* In this study there should be no consideration of problems peculiar to small groups in our population because these problems are too numerous for consideration in all schools. They should be problems which are characteristic of America as a whole, and which when solved would bring benefits to large numbers of our population.
3. *It should be indicative of trends, of conditions of vital importance to our present-day society.* Little emphasis will be given to problems or generalizations which might develop historical foundations, however important they might be for study. This study is more concerned with the pressing, vital problems of today.
4. *It should be of importance to more effective living in our modern society (general) as contrasted with making a living (vocational).* This simply means that such problems as the selection of seed for planting or the building of barns would not be included. These would be useful in the development of materials for training our future farmers in the technical jobs necessary to good farming (vocational education). But this would include those problems and generalizations which are of importance in the education of boys and girls for more effective living in our present society (general education).

All problems and generalizations had to satisfy the foregoing criteria; but certain other criteria were added which the mass of problems and generalizations should satisfy. That is, individual problems or generalizations might satisfy only parts of the following criteria, but collectively should satisfy all. Taken as a group, the problems and generalizations should satisfy these criteria:

1. *They should represent the social, economic, and political interests of our people.* These categories are used because they represent a rather commonly found division of source materials for use in schools. They also include the principal interests of people.
2. *They should express the more important relationships between agriculture and our other American industries.* Many problems in living are caused by the interaction, by the interdependence,

of our several industries. Here there is concern for problems which agriculture may cause through its interaction and interdependence with other industries.

TYPES OF PRINTED SOURCE MATERIALS SELECTED FOR ANALYSIS

Various types of source materials were examined in a preliminary period of reading to determine which would probably be most fruitful of materials dealing with the desired types of agricultural problems and generalizations. Materials which were used in previous social science curriculum studies were examined. *Since it would be virtually impossible for an individual to analyze all published materials relating to agricultural problems, a sampling of three main sources was decided upon.* As stated previously (see page 6) current magazines, books, and national reports were found to be the best source materials for analysis.

SELECTING MAGAZINES

There are many magazines which publish articles dealing with agricultural problems and there would tend to be much duplication of materials in the different magazines. Some would be more productive of the desired type of articles than others. Therefore, in order to makc a preliminary selection of magazines, the Reader's Guide was consulted and a frequency tabulation was made of all agricultural articles appearing in all the magazines there listed during the years 1923 to 1932 inclusive. (The reasons for selecting this ten-year period are given below.) Then the Agricultural Index was consulted and a tabulation was made of all articles (for the same ten-year period) appearing under the following headings: Agriculture, United States; Agriculture, Economic Aspects, and Sociology, Rural. After the tabulations were completed, all magazines which did not have at least fifteen articles published during the ten-year period were ruled out. Then a sampling from each of the remaining magazines was read, and later from other magazines as well. In view of the criteria previously established for the selection of problems and generalizations (see pages 8-9) and as a result of this preliminary reading in magazines, several principles were drawn up for the final selection of the magazines. The following principles were applied in selecting each magazine:

1. *It should be published throughout the ten-year period of 1923–1932.* In order to select materials presenting thought on our contemporary agricultural problems, a ten-year period was used because it would be possible to gain a picture of recent trends in such a period of time, and because ten years would give a wider sampling of materials than a lesser number. Also, during this ten-year period more articles dealing with agriculture in its national phases appeared than during any other recent ten-year period, up to the year 1932. Beginning with the year 1933 there has been a wide variety of national agricultural legislation and numerous articles dealing with our agricultural problems. However, the same fundamental problems existed in 1932 that exist today, except for some variation in intensity or importance. It would be interesting to include a study of recent literature on agricultural problems, particularly since the advent of "New Deal" legislation, but such is not within the province of this study.
2. *It should be productive of many articles dealing with American agricultural problems over this ten-year period.* Magazines which published agricultural articles throughout the ten-year period would be preferred to those publishing articles only occasionally. Magazines which published the largest number of articles would naturally be selected on this principle. As a result of the preliminary tabulation of articles recorded in the Reader's Guide and the Agricultural Index, magazines publishing fewer than fifteen articles during the ten-year period were not included at first.
3. *It should be published in the United States of America.* Occasionally there are foreign interpretations of our American agricultural problems, but this study is not so much concerned with them as it is with the critical interpretations of our own notable interpreters. These can best give the different American interpretations of American agricultural problems.
4. *It should be productive of critiques on the agricultural industry,*

not merely of news concerning the industry. There is greater promise in views, in ideas, in criticisms than in news, when looking for problems and generalizations.

5. *It should have a wide (geographical) circulation over America, but not necessarily a large (number) circulation.* This would tend to require a national viewpoint in editing the magazine, and problems of one particular section would not be emphasized all the time. Such a principle would exclude most state publications.

As in the case of the criteria for the selection of problems and generalizations (page 8), likewise with these principles for the selection of magazines it was necessary to establish certain additional criteria or principles. The five principles just noted were applied in the selection of individual magazines, while the following principles were applied to the group of magazines:

1. *They should represent the views not only of the agricultural interests but of other interests as well.* In a consideration of agricultural problems it is well to see (distinguish, understand) the views of business leaders as well as farm leaders, of manufacturers as well as producers, of consumers as well as producers and distributors. Thus there would be both rural and urban views of agricultural problems. There are differences of viewpoint to consider in the solution of society's problems, especially those relating to agriculture.
2. *They should represent the views of both students and workers in the agricultural industry.* Both the "theoretical" and the "practical" interpretations of our agricultural problems are needed, since both enter into their definition and solution, on a national scale.
3. *They should represent the social, economic, and political interests of our people.* This principle is included so as to correspond with criterion No. 1 explained on page 9. Accordingly there should be some magazines dealing primarily with social problems, some with economic problems, and some with political problems.

THE MAGAZINES SELECTED FOR ANALYSIS

When the foregoing principles were applied to the total list of magazines, there were relatively few magazines left for analysis. These, however, contained a large number of articles on agriculture. The variety of interests represented can be easily recognized from the titles of many of the magazines finally selected. The final list was as follows:

Agricultural Engineering, Agricultural Review, American Economic Review, American Journal of Sociology, Annals of the American Academy of Political Science, Bureau Farmer, Country Gentleman, Current History, Journal of Farm Economics, Journal of Political Economy, Literary Digest, Nation, Nation's Business, New Republic, Outlook, Political Science Quarterly, Review of Reviews, Rural America, Saturday Evening Post, Wallace's Farmer, and *World's Work.*

SELECTION OF ARTICLES IN THE SEVERAL MAGAZINES

Although the selection of magazines was based partly on the total number of articles on agriculture published in them, not all these articles were pertinent to this study; hence the need for some basis of determining particular articles to be analyzed for problems and generalizations. Of the various topics noted in both the Reader's Guide and the Agricultural Index, the following seemed most productive of the needed materials: Agriculture, United States; Agriculture, Economic Aspects; and Sociology, Rural. So the titles of all articles noted in the Agricultural Index, under the topics just stated, were copied first. Next, the titles of all articles noted in the Reader's Guide under the same headings were copied. Duplications in articles secured from the Guide and the Index were eliminated and the final list was written in a permanent register where each article was given a number. This register furnished a means of checking on the articles read, of noting the value of the articles, and of arranging articles for the economical use of reading time. As a result of this procedure, some seven hundred and fifty magazine articles were selected for analysis. Through this procedure a relatively large sampling of articles was gained, from a variety of magazines, dealing with agricultural problems published over a period of ten years.

SELECTION OF THE BOOKS FOR ANALYSIS

Books were used as a second main source of materials for analysis. In order to help determine what books should be selected for analysis, four different lists of agricultural books were consulted first, but only those books published during the years 1923-1932 were noted. The four lists consulted were: The Book Review Digest, the Agricultural Index, A Bibliography of the History of Agriculture in the United States, and A List of Books for College Libraries. In accordance with the criteria previously established for the analysis of printed materials (see page 8), many of the books noted in the four lists were eliminated on the basis of their titles. For example, a book on feeding farm animals was rejected because it obviously dealt with a technical, vocational aspect of the agricultural industry. Not all decisions could be made so easily; therefore, it was necessary to read at length in some books before rejecting them. After this preliminary rejecting of books, a tabulation was made of the books remaining in the four lists and those appearing in two or more of the lists were placed in a combined list.

This combined list of twenty-six books was then submitted to a jury of fifty men and women who were interested in agricultural problems and who were recognized as authorities in their respective fields of work. The jury selected was composed of deans of colleges of agriculture, professors of agricultural economics and rural sociology, extension leaders in the United States Department of Agriculture, specialists in this same Department, and officers of national farm organizations. The following letter together with the list of twenty-six books was sent to each member of the jury.

Dear ———

I am making a study of the problems of American Agriculture that should be considered in our public schools. In my endeavor to arrive at a statement of these problems, I am consulting several sources. On the enclosed sheet I have made a list of books published in the period of 1923-1932. I wish that you would merely check the ten books which you think would give me the best exposition of American agricultural problems. If my list does not include the best ten books, please indicate on the reverse side of the sheet any additional books which you believe would be profitable for my use. There are several conditions which the selected books should meet, namely:

1. The books should have been published within the ten-year period 1923-1932 inclusive.
2. The books should treat some phase or phases of our agricultural problems from a national viewpoint, not from a strictly sectional or local viewpoint.
3. The books should be written particularly for the lay reader; that is, they should not be college textbooks on agricultural economics, for example.
4. The selection of ten books should give a general picture of the social, economic, and political problems to be found in a study of American Agriculture.
5. The books should not include national studies such as those reported by the Industrial Conference Board, for example.

Kindly make this selection of books on the basis of your own reading and return the enclosed sheet in the addressed envelope enclosed. Your assistance in the selection of these books will be greatly appreciated.

This request has the approval of Professor Jesse H. Newlon of Teachers College, Columbia University, who is the faculty adviser for this study. Dr. Newlon asked me to state that he will appreciate it very greatly if you can co-operate in the compilation of this list of books.

Thirty members of the jury replied to this letter, and the replies were tabulated so as to get a frequency of choice. Of the twenty-six books originally submitted with the letter, eleven received the vote of ten or more judges, so these were selected as the books to be analyzed for problems and generalizations. This list included the following books:

Bailey, Liberty H. *The Harvest of the Year to the Tiller of the Soil.* New York: The Macmillan Company, 1927. 209 p.

Beard, Charles A. and Beard, Mary. *The Rise of American Civilization.* New York: The Macmillan Company, 1930. 380 p.

Black, John Donald. *Agricultural Reform in the United States.* New York: McGraw-Hill Company, 1930. 380 p.

Engberg, R. C. *Industrial Prosperity and the Farmer.* Washington: Brookings Institution, 1927. 286 p.

Kirkpatrick, Ellis Lore. *The Farmer's Standard of Living.* New York: The Century Company, 1929. 299 p.

Lord, R. *Men of the Earth.* New York: Longman, Green & Company, 1931. 298 p.

Sanderson, D. and Others. *Farm Income and Farm Life.* Chicago: University of Chicago Press, 1927. 324 p.

Spillman, W. J. *Balancing the Farm Output.* New York: Orange Judd Publishing Company, 1927. 126 p.

Wallace, Henry C. *Our Debt and Duty to the Farmer.* New York: The Century Company, 1925. 232 p.

Warren, G. F. and Pearson, F. A. *The Agricultural Situation: Economic Effect of Fluctuating Prices.* New York: John Wiley and Sons, 1924. 306 p.

Williams, James Mickel. *Our Rural Heritage: The Social Psychology of Rural Development.* New York: Alfred Knopf Company, 1925. 246 p.

Some few other books were suggested by the jury. No one of these added books received more than three votes from the jury, so none was added to this original list of eleven books. An attempt was made to avoid the inclusion of textbooks in the first list of eleven books, but judging from the replies of the jury several of the books selected were in use as textbooks. The original intention was to include only those books which were written primarily for the lay reader because such books would probably bring knowledge more generalized, and ideas more understandable for the "layman" than would technical textbooks; for the purpose was to secure materials which would be useful to urban as well as rural schools. However, so many of the jury suggested the inclusion of some textbooks on the social and economic problems of agriculture that some were included. Another jury was asked for suggestions as to appropriate textbooks for the presentation of agricultural social and economic problems to secondary school pupils. The replies indicated the advisability of including for analysis the following additional books:

Gray, Lewis C. *Introduction to Agricultural Economics.* New York: The Macmillan Company, 556 p.

Sims, Newell LeRoy. *Elements of Rural Sociology.* New York: Thos. Y. Crowell Company, 1928. 698 p.

Taylor, Carl C. *Rural Sociology.* New York: Harper & Brothers, 1926. 511 p.

Taylor, Henry C. *Outlines of Agricultural Economics.* New York: The Macmillan Company, 1925. 610 p.

In this manner, then, fifteen books were selected for analysis.

SELECTION OF THE REPORTS FOR ANALYSIS

The third main source of materials included for analysis was national reports that were published during the same ten-year period of 1923-1932. Since the number of reports was so limited, all that could be located were included for analysis. They were:

Report of the President's Agricultural Commission, 1924.

The Agricultural Problem in the United States. The National Industrial Conference Board, 1926.

The Condition of Agriculture in the United States and Measures for Its Improvement: A Report by the Business Men's Commission on Agriculture, 1927.

Reports on the Agricultural Situation by the Special Committee of Land Grant Colleges and Universities, Chicago, 1927, 1932.

Report of the Committee on Recent Economic Changes of the President's Conference on Unemployment, 1929.

Report of the Conference on Economic Policy for American Agriculture, University of Chicago, 1932.

In addition to the above reports, *Rural Social Trends* by Brunner and Kolb was read. This book, published in 1933, could not be included in the list for analysis because of the date of publication, but the analysis was checked against the problems and generalizations there cited.

So the completed list of source materials for analysis included seven hundred and fifty magazine articles, fifteen books, and seven national reports, all published in the ten-year period 1923-1932.

THE TECHNIQUE OF ANALYSIS USED

As stated on page 6, the technique of analysis used in this study was somewhat similar to that followed in earlier studies.

Some of it has already been described in this chapter, but some of it remains to be explained. The magazine articles were the first materials to be analyzed for problems and generalizations. Later the books and the reports were read alternately.

All the magazine articles and other materials were read in full. At first it was thought that it would be possible to read only the first and the last sentences of paragraphs, and there find the problems and generalizations. But trials with both types of reading showed the necessity for reading all the materials.

The titles of all magazine articles to be read were placed in a register according to the name of the magazine (for example, all articles found in the *Country Gentleman* were grouped together), and in chronological order of publication (so that articles could be readily located by years and volumes). Each article was given a number in the register.

As the several articles were read, the problems and generalizations were written down on ruled paper. If the statement was a problem, the letter P preceded it; if the statement was a generalization, the letter G preceded it. The number of the article (given in the magazine register) was also placed with the preceding letter so that the source of any problem or generalization might be easily found. For example:

100 P	"Should government lands be rented for the production of wheat when there is already an overproduction of wheat?"

This indicated a problem from magazine article No. 100.

100 G	"Through co-operatives, small farms are benefited in ways supposedly available only to great corporations."

This indicated a generalization taken from the same magazine article No. 100. In like manner each book and each national report cited on pages 15-17 was given a key number, and all problems and generalizations found in them were recorded in the same way as those found in the analysis of the magazine articles.

After all the source materials had been analyzed and the problems and generalizations recorded, the next step was to organize the latter into a more serviceable form for curriculum workers.

Since all problems and generalizations were recorded on ruled paper, it was necessary to cut this original paper horizontally into separate strips with a problem or a generalization on each strip. Then all the problems were placed in one group and the generalizations in another. The problems were then separated into political, social, and economic sub-groups. Next, the generalizations were read and those applying to any particular problem were placed with that problem. After all the generalizations were re-read, it was quite apparent that they could not all be grouped under the stated problems from the source materials. So problems were framed that seemed to be answered by different groups of generalizations. Thus the list of problems includes those which were stated directly in the source materials and those which were implied by a study of the generalizations.

When the problems were brought together, it was found that there were many statements of the same problem; hence a restatement was made and several problems were included in one statement. Where several hundred problems were secured from the source materials an attempt was made to combine statements without the loss of the essential meaning, and so the number of problems in the final list was reduced to fifty-three. In like manner it was necessary to combine generalizations with similar meaning and reduce the number from several thousand to several hundred. In all these combinations of statements an effort was made to preserve the essential meaning and yet simplify the wording as much as possible. It was necessary to combine and recombine statements many times before the final form of organization was accepted. The greatest difficulty was found in reducing an almost overwhelming number of statements to a relatively small number of simply worded problems and generalizations. Such problems and generalizations follow in Chapter III.

CHAPTER III

SOME IMPORTANT PROBLEMS OF AMERICAN AGRICULTURE TOGETHER WITH RELATED GENERALIZATIONS

THE purpose of this study (as related in Chapter I) is to define and state some important American agricultural problems together with their related generalizations in a manner that will be helpful in the construction of modern curricula for both urban and rural schools. In this chapter these problems and their related generalizations are stated. Partly for greater ease of understanding, the problems are organized into three main groups, namely: Some Important Political Problems of American Agriculture; Some Important Social Problems of American Agriculture; and Some Important Economic Problems of American Agriculture. These three main groups are further subdivided into Centers for Units of Study, such as "The State and Local Government" and "Land." These centers are focal points that may be of help to teachers in developing titles for units, and in discovering interesting agricultural curriculum areas. The points of unity (the centers) are in reality big concepts that are often found in discussions on politics, sociology, or economics. The generalizations grouped under a given center for units are all related to a big concept, such as "The National Government." Furthermore the problems grouped under a given center for units are all closely related to that center. And generalizations found with a given problem are closely related to a solution of it.

This chapter is divided into two parts. Part I is simply the organization of the agricultural problems according to the arrangement stated in the preceding paragraph, while Part II is the statement of the same agricultural problems together with their related generalizations.

PART I

THE STATEMENT OF THE AGRICULTURAL PROBLEMS

Division One of the Problems

Some Important Political Problems of American Agriculture

CENTERS FOR UNITS OF STUDY

A. The State and Local Government

Problems: Are there any changes needed in rural government? How shall we reorganize our rural government so as to make it function more effectively? What are some of the defects in our support of rural government? What shall be the bases for the support of rural government?

B. The National Government

Problems: What should be the general function of government with respect to the agricultural industry? What should be the tariff policy in relation to agriculture? To what extent should our public agencies enter into the fixing of prices on agricultural products? To what extent should our public agencies enter into the distribution or marketing of agricultural products? To what extent should our public agencies enter into the control of production?

Division Two of the Problems

Some Important Social Problems of American Agriculture

CENTERS FOR UNITS OF STUDY

A. Educational Agencies

Problems: What are some pertinent characteristics of rural schools? What unit of administrative control should we have for our rural schools? How should we support our rural public schools? What are some rural educational agencies other than the schools? What is the status of recreation and leisure time among rural people?

B. Other Social Agencies

Problems: What is the health status of rural people? What agencies are helping to safeguard the health of rural people? What are some pertinent characteristics of the present church in rural areas?

What are some significant needs and developments in the rural church? What emphasis is attached to other rural organizations?

c. Rural Population

Problems: What are some principal characteristics of rural people? What are some distinguishing characteristics of the rural population in contrast with the urban population? What should be the general relationship between rural and urban people? How much of our total population should be engaged in the agricultural industry? What recent movements or changes in population have taken place in rural areas? How shall we conceive the standard of living in rural areas? How do land ownership and land tenantry affect rural life?

Division Three of the Problems

Some Important Economic Problems of American Agriculture

a. Land

Problems: What are some general considerations in the utilization of our land resources? To what different uses should our land be put? What should be our national policy with relation to the development and settlement of our agricultural lands? What disposition should be made of the submarginal agricultural land? How can we maintain the productive powers of our agricultural lands? What causes our farm lands to change in value?

b. Production

Problems: How can we balance our agricultural production? To what markets shall we adjust our agricultural production?

c. Distribution

Problems: What are some of the most notable defects of our distribution of agricultural products? How shall we distribute agricultural products to consumers? What are some of the principal effects of modern transportation on the agricultural industry?

d. Prices

Problems: What importance is attached to the price of agricultural products? What causes increase or decrease in agricultural prices? What are the effects of increased or decreased prices, of inflation or deflation, on agriculture? What effect does the world price of agricultural products have on the domestic price? How can we more nearly stabilize the price levels for agricultural products?

E. Credit and Income

Problems: What are the credit needs of agriculture? How should agriculture share in the national income?

F. Related General Economic Factors

Problems: What is the general economic pattern in American agriculture? What is the economic relationship between agriculture and our other great national industries? What is the influence of costs in the agricultural industry? How has the factor efficiency influenced the agricultural industry? How has machine development affected the agricultural industry? How is labor employed in the agricultural industry? What type of farming shall we have? What kind of farm organization should we have in America?

PART II

THE STATEMENT OF THE AGRICULTURAL PROBLEMS AND RELATED GENERALIZATIONS

Division One of the Problems and Generalizations

Some Important Political Problems of American Agriculture

CENTERS FOR UNITS OF STUDY

A. The State and Local Government

Problem: Are there any changes needed in rural government?

Related Generalizations:

1. There are too many small local units, too many overlapping and duplicating agencies, too many administrative areas which are now inadequate in size and resources to perform the services required of them.
2. Of all the institutions with which rural society is concerned, local government has probably changed least with respect to its units and forms.
   a. Underlying the carry over of the traditional governmental units and forms is the notion that there must be uniformity of the local governmental structure in a given state.
   b. Natural inertia and a misguided effort to preserve for local authority the things which are truly local, have led to a blind opposition to all changes.

Problem: How shall we reorganize our rural government so as to make it function more effectively?

Related Generalizations:

1. The field of governmental services should be re-examined and services reallocated in the light of modern needs.
   a. State and interstate planning is of first importance.
2. States have been steadily acquiring greater influence over and control of local functions and agencies; and in most instances without complete changes in the framework of existing governmental units.
3. County boundary lines should be changed so as to create governmental units that more nearly coincide with natural social and economic units, regard being paid for such factors as population centers and highways.
   a. Farm people are situated on the slopes of social-economic watersheds draining into villages or cities, and these watersheds ignore township and county lines.
   b. The farmer always has had more interest in state and national government than in local government.
4. Both urban and rural people should have opportunity for the enjoyment of social and economic services, which they help create, but which they are often denied because of existing governmental boundaries.

Problem: What are some of the defects in our tax support of rural government?

Related Generalizations:

1. Our tax system absorbs a relatively high proportion of the income of agriculture.
   a. Even in the short run, the cost of taxes is shifted so that determining the actual burden borne by a given social class is very difficult.
2. Our local governmental agencies have spent too great a proportion of their incomes on functions which are essentially state and national in character.
3. The tax base for the support of rural government has been narrow.
   a. Rural taxes are chiefly general property taxes, levied on the basis of capital value.
   b. The property tax has resolved itself, through the inability to find the intangible property, into a real estate tax.

Problem: What shall be the bases for the tax support of rural government?

Related Generalizations:

1. We need a system of taxes which makes the levies on several different bases, which distributes the burden more equally, and which makes it easier to secure adequate revenues.
   a. Wealth is often concentrated in small geographic areas, but is derived from much larger areas.
   b. Any concentration of wealth in one place, such as in a mine or in a large city, should be a source of income upon which counties, states, and even the nation should be able to draw.
   c. Net income is a practical measure of ability to pay, and every citizen having taxable capacity should contribute to the support of the government under which he lives and from which he derives daily benefits.
2. The redistribution of the tax burden between state and local governmental units would help greatly in lessening many of the inequalities in the present situation.
   a. The states should, so far as practicable, relinquish the general property tax to the local governmental units and obtain such revenues as they need through income taxes, excises, and business taxes.
   b. Some principal objects of local governmental expenditure, such as schools and highways, are of state and national importance and should receive a greater proportion of the total support from the latter governments.
3. The tax system should be so modified as to permit the taxation of agricultural land in keeping with its production capacity.
   a. Unless all rural taxable property is assessed relatively on the same basis, there can be no equality of taxation.
4. Fundamental tax revision cannot properly be viewed merely from the standpoint of agriculture alone, because it concerns all economic interests.
   a. In all questions of readjustment of the tax burden on agriculture, there is a need for constructive co-operation between farm organization and organizations representing other economic interests in each state.

CENTERS FOR UNITS OF STUDY

B. The National Government

Problem: What should be the general function of government with respect to the agricultural industry?

Related Generalizations:

1. It should be the function of government to formulate national policy and to provide, through national and state legislation, a favorable basis upon which a sound and vigorous agriculture may be maintained.
2. The primary requisites of people are the supplies of food, clothing, and shelter. To safeguard the opportunity to obtain these is the first responsibility.
   a. Drought-stricken areas or other stricken areas have a right to governmental appropriations for their relief.
   b. Because of the unavoidable hazards involved in farming and the vital relationship of farming to every food consumer and to many industries, we may justify governmental effort to safeguard agriculture.
3. That the state has moved far toward social control of the industry is evident from even a cursory view of its extensive and multifarious farm activities.
   a. The idea that governmental and economic activities occupy water-tight compartments with no connection is erroneous.
   b. Agriculture must rely on the government for research activities and similar benefits which are left to private initiative in other industries.
   c. The American farmer is in need of much information from governmental agencies in order that he may intelligently develop his producing and marketing program.
4. Our economic unit now is the nation, hence we are justified in supplementing local revenue with state and national revenue for the benefit of agriculture.
   a. It is a function of government to help alleviate the hardships of unemployment in agriculture as in other industries.
5. Economic legislation affecting urban industries should be enacted with a consideration of its probable effect on rural as well as urban industry.
6. Improvement in farming, if it is to be permanent, must come in a large measure through the individual efforts of sane, thoughtful, progressive farmers rather than through governmental action.

Problem: What should be the tariff policy in relation to agriculture?

Related Generalizations:

1. American agriculture has been largely a surplus product industry and, therefore, not in a position to profit from tariff protection.
   a. The general effect of tariff barriers is to reduce the volume of both imports and exports.

2. The effects of the tariff are often over-emphasized by both its supporters and its opponents.
3. A tariff on agricultural products should be an integral part of a national policy.
   a. Agriculture should have the same protection through our tariff as other industries enjoy.
   b. Rates of duty on foreign-grown farm commodities which seek markets in our country must be adequate to permit our farmers to enjoy that profit which guarantees the American standard of living.
   c. Tariff schedules should be revised so as to give American farmers full benefits of the American market for all agricultural products produced in this country, and for all articles manufactured from such products.

Problem: To what extent should our public agencies enter into the fixing of prices on agricultural products?

Related Generalizations:

1. Price-fixing, direct or indirect, belongs to the class of governmental acts which are economically questionable.
   a. The biggest difficulty in price-fixing is to get the facts needed as the basis for action.
   b. Government price-fixing is not workable unless both production and consumption are at the same time rigidly controlled by the government.
   c. The Government itself cannot enter into the business of fixing prices without objection from several industrial sources, but there is little objection to the same thing being done by private corporations with funds borrowed from the Government.
   d. Between fixing value in terms of money and fixing value in terms of commodity exchange, there is little real difference.
2. Some of the large urban industries using farm products as raw materials are able to control to a large extent the prices paid for such raw materials.
3. What agriculture asks is that it be given the same right as other industries to name a price, in the first instance, and that it be enabled to acquire an organization which will secure the power to maintain that price.
   a. No agricultural organization can fix the price at which the product is to be sold and sell much of its product unless that price also suits the consumer.
4. Agricultural legislation, both national and state, is necessary as an aid in minimizing fluctuations in the price of farm products.

Problem: To what extent should our public agencies enter into the distribution or marketing of agricultural products?

Related Generalizations:

1. In this modern and complex organization of our economic and social life, there must be some public regulation of marketing farm products.
   a. Integration of market control is not possible through any other agency than the national government.
2. While the Government should not engage in buying and selling farm products, except in emergencies, it should take a part in finding markets for our agricultural surpluses.
   a. One of the most important opportunities for public agencies is to carry on research which will give agriculture the necessary information for orderly marketing.

Problem: To what extent should our public agencies enter into the control of production?

Related Generalizations:

1. The individual farmer should be free to plant whatever acreage of any crop that he pleases or keep whatever livestock that he pleases.
2. Some agency of the central government should exercise the power to control production of agricultural products.
   a. Neither county programs of production, nor state programs of production, are worth anything so long as they are not co-ordinated with a national program of production.
   b. The Government should assume no responsibility in normal times for crop surpluses clearly due to over-extended acreages.
   c. Our public agencies, including state agricultural colleges, have tried to teach the farmer increased productivity, but he thinks in terms of increased gross production.

Division Two of the Problems and Generalizations

Some Important Social Problems of American Agriculture

CENTERS FOR UNITS OF STUDY

A. Educational Agencies

Problem: What are some pertinent characteristics of rural schools?

Related Generalizations:

1. Rural schools, although cast in the midst of the circumstances of

farming and farm life, are too often more adapted to urban conditions than to rural conditions.

   a. Little consideration is given in most rural elementary schools to the fundamentals of rural life, and the course of study seems to be shaped to meet the requirements for entrance into the high school or college.
   b. Just as the rural school is an institution, so is it largely a set of institutionalized courses of study.
   c. The influx of country youth into village schools necessitates a re-thinking of the curriculum in terms of the needs of the two groups, rural and town.

2. Rural school opportunities are limited as compared with urban school opportunities.
   a. Good rural communities generally provide good school opportunities.
   b. Because the unit of rural school support and administration is so generally local, rural school opportunities differ widely in different districts.
   c. The rural school is more poorly taught than the city school.
   d. The rural school term is shorter than the city school term.
   e. Rural school attendance generally falls below that of city school attendance.
3. The problems of rural education are much greater in scope and magnitude than merely the problems of the rural school.
   a. Farming, inherently, has more educational possibilities for children than most other vocations.

Problem: What unit of administrative control should we have for our rural schools?

Related Generalizations:

1. The single-room country school is in disfavor with educators.
2. The trend toward school consolidation is clearly indicated.
   a. The purpose of school consolidation is to furnish better schools and better education for rural children.
3. There is a general tendency in the direction of centering rural education in villages and towns, whether schools are consolidated or not.

Problem: How should we support our rural public schools?

Related Generalizations:

1. State aid and national aid should be given our rural schools where needed to provide adequate schools.
   a. Urban wealth should help support rural schools on the principle

that the total social wealth must be utilized for the total social health.

b. An excessive proportion of the financial burden of giving our total population the high school and elementary school education has fallen on farm parents.

c. One of the trends is the increased proportion of the rural school budget coming from sources outside the school district itself.

2. There is an increased demand on the part of rural people for schools somewhat similar to those at the disposal of urban people.

   a. Judged by capital investment, budget, and number of persons employed, the school is the most important activity of local government in rural America.

Problem: What are some rural educational agencies other than the schools?

Related Generalizations:

1. The agricultural extension services of national and state governments are of great importance in getting information to the farmer.
2. Farm organizations are giving increased recognition to the importance of educational agencies as aids in working out the solution of agricultural problems.
   a. The building and carrying on of a soundly organized agricultural co-operative constitutes a definite contribution to rural adult education.
   b. The organization of agriculture in small and widely separated units makes it very difficult for the industry to carry on its own research and educational activities.
3. Some additional educational agencies of importance to agriculture are: The home, the church, the theater, the radio, the press, and national farm organizations such as the Farm Bureau, the Farmers' Union, and the Grange.

Problem: What is the status of recreation and leisure time among rural people?

Related Generalizations:

1. The leisure time of farm people is directly conditioned by the labor demands of the farming enterprise.
2. The opportunities for recreative activity, on the average farm, are rather limited.
3. Rural people who formerly went to their neighbors for recreation now often go to the village and city.

4. Opportunity for constructive use of leisure time can not be provided under the isolated farm situation unless there is organized and diligent effort to create it.
   a. Some tangible means of improving recreation are rural parks, playgrounds, and community buildings.
   b. The most universal rural play program is developed in connection with the schools.

CENTERS FOR UNITS OF STUDY

B. Other Social Agencies

Problem: What is the health status of rural people?

Related Generalizations:

1. The health of the farm family is less adequately safeguarded by organizations and by the enforcement of health laws than is the health of the urban family.
   a. The great cost of disease in rural districts is due largely to the inadequacy and scarcity of existing medical services and hospital facilities.
2. Insanitary conditions are widespread in rural areas.
   a. General insanitary rural conditions are caused by: the individualistic mode of farm life; ignorance of the nature of disease and its dissemination, and methods of protection against it; diffusiveness (scattered population); and expense.
   b. Insanitary rural conditions affect not only rural dwellers but urban as well.

Problem: What agencies are helping to safeguard the health of rural people?

Related Generalizations:

1. The extension of hospital service to rural areas has been one of the most recent developments.
2. Rural public health associations have recently made outstanding contributions to rural health.
   a. Maintenance of health is becoming each year more of a social responsibility in rural areas.
3. The public health nurse is of great importance in the public rural health program.
   a. There has been a decided growth in the health demonstration method in rural areas.

Problem: What are some pertinent characteristics of the present church in rural areas?

Related Generalizations:

1. Because widely different conditions prevail throughout the country no general description of the rural church can do it justice.
2. The rural church is failing to furnish the ethical and moral leadership needed by rural communities.
   a. The leadership of rural churches gives little evidence of response to changing tendencies.
3. Pastors of rural churches are poorly trained, poorly paid, and restless.
4. Rural church buildings and equipment in the open country are generally meager and in need of repair.
5. Limited financial support is given most rural churches.
6. An increasing proportion of village church membership is resident in the country.
   a. The rural church is deficient in membership.
7. Rural church attendance is almost universally poor.

Problem: What are some significant needs and developments in the rural church?

Related Generalizations:

1. The rural church needs to develop a more aggressive, progressive social program.
   a. The needs to which the rural church is called to minister are no longer altogether personal; they are also communal.
   b. Changing rural conditions require, but have not yet secured, a corresponding church institutional readjustment.
2. The trend toward the union or community rural church is nationwide.
   a. Adjustments involving more co-operation among rural churches seem apparent; yet evidence of competition persists.

Problem: What emphasis is attached to other rural organizations?

Related Generalizations:

1. In giving consideration to agricultural difficulties, a popular conclusion to arrive at has been that the key to the situation is organization.
   a. Effective rural organizations are community grown, not manufactured importations.
2. Agriculture has inherent difficulties which cannot be overcome by the individual farmer acting alone.
   a. Rural people are generally less able to combine for action and are more helpless in making their demands effective than urban people.

3. The fullest utilization of existing organizations before new ones are essayed should be regarded.
   a. The family is generally recognized as the unit of most organizations in the country.
   b. Hard times on the farm always stimulate the growth of farmers' organizations.

CENTERS FOR UNITS OF STUDY

c. Rural Population

Problem: What are some principal characteristics of rural people?

Related Generalizations:

1. The juristic attitude of respect for law was an outstanding attitude in the early neighborhood and it has remained fundamental to this day in rural areas.
2. The farmer has been a national stabilizer.
   a. The farmer is generally slow to change his opinions.
   b. Adherence to custom has characterized rural communities all over the world.
3. Ideas are subordinate to attitudes in the rural mind.
4. The individualism of the farmer affords a suggestive contrast with the growing reliance of agriculture upon public regulation and state aid.
5. The conditions of its existence have fostered an unusual degree of interdependence among the members of the farm family.
6. The farmer is a business man bound by the laws which operate in other business fields.
7. There is a love of the land which is rarely bred out of the rural people.

Problem: What are some distinguishing characteristics of the rural population in contrast with the urban population?

Related Generalizations:

1. The home life of the farm family is interwoven with the operation of the farm in a way that is not possible in any other line of business.
2. The integrity of the family is more likely to be understood and supported among farmers than among urban folk.
   a. A man in the country is more likely to utilize the economic value of women and children than is the city man.
   b. Family continuity is stronger in agriculture, for a man does not learn his land and climate and the possibilities of his place in much less than a working lifetime.

3. There is a far larger natural increase in the rural than in the urban population.
   a. There are more married people living in rural areas than in urban areas, in proportion to population.
   b. The farm population has proportionately more children, fewer older people, and more males than the city.
4. On the whole, there is greater homogeneity in the population in the country than in the city, except in the age and race factors.
5. The rural non-farm population is made up of diverse groups with sharply different economic backgrounds and bound together by no such major occupational interest as is the rural farm group.

Problem: What should be the general relationship between rural and urban people?

Related Generalizations:

1. Persons in congregated centers need always to be reminded of the dispersed and scattered places.
   a. Most of the prejudice existing between country and city folk is due to misunderstanding.
2. Our national life has become very complex, and there is such an interdependence of one group of citizens upon the other that the welfare of all groups must be taken into consideration in dealing with the problems of any single group.
   a. The farmer can never make his greatest contribution to our national life until he and the city man become aware of their organic relationship, when each comes to realize that the other has something of importance to contribute.
3. The interrelations of the agricultural village and the open country population are close and the contacts implicit in these interrelationships are increasing.

Problem: How much of our total population should be engaged in the agricultural industry?

Related Generalizations:

1. There is clear evidence of occupational overpopulation in the agricultural industry.
   a. The number of producers in any field is now of utmost significance.
   b. A decreasing proportion of the population is being required to supply the world with food.
   c. So far as a back-to-the-land movement proposes to create new full-time farmers, it sets itself in opposition to a major economic trend.

   d. Agencies which tend to hold an excess of farmers on the land should be displaced by agencies to facilitate the movement of the surplus farm population into other occupations in order to maintain a proper balance between rural and urban population.
2. The preservation of the agricultural industry on a basis which will maintain on the land a rural population that may continue to contribute to the public welfare something more than adequate food supply is essential to the permanent well-being of this country.
   a. The elaboration of civilization depends on the proportion of population that can be spared from the production of food and other raw materials contributed by agriculture.
3. Partly because of a greater natural increase in population in rural districts, there has been a fairly constant need for movement of people to cities.
4. Practically all long-settled rural areas tend to decline in population.
   a. New people must go into the agricultural industry from time to time, just as in any other industry; otherwise, decay is inevitable.

Problem: What recent movements or changes in population have taken place in rural areas?

Related Generalizations:

1. The mobility of America's population has been one of the important influences producing changes in the composition and structure of the population, both urban and rural.
2. There is a constant sorting of men, according to their abilities, so that they may occupy the kind and type of land which not only suits their wishes but also their respective financial capacities for organization and management.
3. Any policy with respect to immigration as it affects our agriculture is a policy with respect to our farm population.
4. Cityward drift of rural people has been one of the most significant of all movements influencing country life.
   a. It has been largely the young men and women who leave the rural districts.
   b. An improvement in farm income has been accompanied by a smaller movement of population from farms and a decline in income has been accompanied by an increased shift to industrial centers.
   c. There has been a tendency toward village growth and a tendency toward a decreasing rate of change, or greater relative stability, in numbers of villages.

d. The cause for the movement from farms to rural villages appears to be not so much economic as social.

Problem: How shall we conceive the standard of living in rural areas?

Related Generalizations:

1. The ultimate ideal in farming is to obtain a return which will enable farmers to build up a satisfactory rural life.
   a. No farmer should be expected to labor without financial reward or to live without a fair share of productive leisure.
   b. Current income is a better index than is accumulated wealth of the material welfare of farmers.
2. For the improvement of rural life, the growth of well-planned and rounded agricultural communities is necessary.
   a. State planning commissions may assist in establishing rural communities which will make accessible the services needed to maintain a modern standard for rural living.
   b. Many of the elements of a higher standard of living must come through community effort and public expenditures.
3. The demands and interests of rural people are becoming manifestly more general, more like urban demands and interests, each year.
   a. On the farm there is little enthusiasm for the live-at-home program.
   b. Essentially in order to get the maximum values from farming and from living in the farming community, every family should have access to the goods, facilities, and services the efficient use of which will insure both the physical and the spiritual well-being of its members.
   c. There is warrant for believing that differences in the scale of living, together with many other differentials between the masses of rural and urban people, are constantly decreasing.
   d. Comparisons between rural and urban scales of living are difficult to make, because rural and urban values are not the same.
   e. The higher the standard of living of the farm people, the larger the percentage of the income that is spent for goods of the sort that are manufactured in the city.
4. The successful farm life is a complex of social and economic balances.
5. One of the most significant trends of the past decade has been the increased recognition and attention given to the family-living aspects of rural life.

   a. Sunshine, music, and labor-saving devices must find their way into all our better farm homes.
   b. The new wants created by the machine age are causing farmers to raise their standards of living.
6. Any progress for building a finer rural civilization must center about the securing of better incomes for farm people and the utilization of incomes in securing the basic conditions of a healthful, wholesome, cultured life.
   a. As income rises among farmers, even as among other workers, more is devoted to culture and advancement.
   b. It is astonishing what a low standard of living a farm family will maintain.
   c. It is the factor of instability of income which operates so strongly against a rising standard of family living in rural communities.
7. Amount of surplus, not production, will determine the farm standard of living.
   a. We have too many farms where the farmer's surplus is represented simply by what he and his family get along without.
   b. The perennial loss of surplus wealth is one of the outstanding causes of a barren country life.
8. Many men honestly believe that most of agriculture's troubles are due to the fact that the farmer's standards of living are too high.
9. If farm communities constantly produce much and constantly consume little, they tend to lower their own standard of living instead of raising it, and the benefits of the farmers' hard work and thrift go to the consumers of the farmers' products (that is, to urban communities).
10. Most farm families find a wealth of satisfactions or values outside the realm of income.
    a. The satisfactions or values of life accruing from the prevailing standard of living depend upon the efficient use of the time as well as the efficient use of goods by the different persons composing the family.
11. The number of persons and the different ages of persons composing the family have direct bearing upon the prevailing standards of living.
12. The house in which a family lives is a fairly good index to the whole standard of living.
13. If we are to have a permanent and balanced agriculture, it must be built on a basis that not only will satisfy the present farming generation but will fulfill the needs and ideals of those who now are young.

Problem: How do land ownership and land tenantry affect rural life?

Related Generalizations:

1. Gifts, inheritance, and savings, aided by a good credit system, are the most important means of acquiring the ownership of land.
2. In nearly all newly-settled regions of good land in the United States there has been a period when practically all farmers were landowners, because the national policy of disposing of public land made it exceedingly easy to acquire a farm.
3. No incentive has yet been found by rural people that will stimulate exertion and thrift like the prospect of acquiring the ownership of a good farm.
4. The better the land, the more tenaciously the first owners and their heirs will hold it for investment long after they have ceased to work it themselves.
   a. Concentration in the ownership of good farm land often leads to popular dissatisfaction and unrest.
5. Landlords in the United States are, as a rule, residents of the district in which their lands are located.
   a. The owners of rented land are quite generally older farmers who have retired or who have more land than they wish to farm.
6. Owners appear to have higher prevailing standards of living than do tenants.
   a. As a general thing, the intellectual interests of farm owners are greater than those of their tenant neighbors.
   b. Farm owners universally have a greater volume and greater diversity of reading materials in their homes than do farmers of any other tenure status.
   c. Now that agriculture has become largely commercial, the basis of the farmers' welfare and independence is no longer land ownership but income.
7. The difference between types of farm tenants is so great that any broad generalizations of needed action in their behalf is difficult of application to all types.
8. Studies in the efficiency of white tenant farming compared with that of owner farming do not indicate a difference as wide or as important as is generally attached to this question.
9. Farm tenancy is an institution which has often provided for getting the land into the hands of those who are in a position to cultivate it, but who are unable to buy farms.
10. Socially considered, it is often desirable that beginners operate

farms belonging to those who for some reason are not in a position to do farming themselves.

11. As land increases in value an increasing percentage of it will be farmed by tenants, because they can better afford to pay rent at a low rate on land value than to burden themselves with the load of debt which would be necessary to purchase it.
12. Tenants in general make as great labor incomes as do owners, or greater, with the notable exception of our southern Negro tenants.
13. Tenancy has been actually and relatively increasing.
    a. The number of tenant families or units is a fair measure of the social-economic influences of tenancy.
14. The general effect of farm tenantry on community life is to create more or less unstable conditions.
    a. The most characteristic feature of tenancy is mobility.
    b. Most of the tenant shifts are from farm to farm within the same community, but many tenants migrate into new agricultural regions.
15. Every phase of institutional life is affected adversely by the prevalence of any large number of tenant families in the community.
    a. The unequal bargaining power of landlord and tenant is one of the most obvious influences in the farm tenantry situation.
16. The tenant almost universally crops the land harder than the owner.
17. Most of the tenancy is a short-term share or cash rent arrangement between owner and operator.
18. A large proportion of tenant farmers are young men.
19. When prices of farm products are high, an unusually large number of hired men become tenants and an unusually large number of tenants buy farms.

Division Three of the Problems and Generalizations

Some Important Economic Problems of American Agriculture

CENTERS FOR UNITS OF STUDY

A. Land

Problem: What are some general considerations in the utilization of our land resources?

Related Generalizations:

1. The basis of all agriculture is the land itself.

2. The ability to support vegetation is inherent in every soil.
3. Farm land varies greatly in its economic capacity and efficiency.
4. The fertility of the soil determines to a large extent the agricultural production.
5. For any class of soils crop yields may be raised or lowered by favorable or adverse treatment.
6. The capital which farmers accumulate is mainly in the form of land, livestock, and their accompanying improvements.

Problem: To what different uses should our land be put?

Related Generalizations:

1. The first step in putting each kind of land to its proper use is a scientific, economic survey of the land.
   a. A given region's available land uses are more likely to be complementary than competitive, and the common interest is best served when all the promising associated uses are encouraged to function to the fullest capacity of our social need.
   b. Land classification for uses involves many political considerations and affects vast business interests.
2. Land economists have repeatedly said that there is too much land in agricultural use.
   a. We have come to the time when we should write a new epic—the epic of adjustments, of regrouping, of retirement from cultivation of lands which the pioneer subdued but which now refuse to yield a reasonable scale of living.
   b. Not all soils and places where farms are located are adapted to farming.
3. A study of land resources and future requirements for farming should be answered in terms of an economic organization pattern; that is, in terms of types of farming in given regions.
   a. Much land formerly in crops should now be put in grass or forests.
   b. The utilization of available farm land resources should be directed between farming uses proper, grazing, and forestry, and should be in keeping with the interests of established farmers and the needs of the nation.
4. Although all classes of economic productive activity require some land, agriculture makes the greatest demands for this agent of production.
5. Regional planning related to water supplies, flood control, residential development, and recreational facilities should be coordinated with the rational use of the land for farm purposes.
   a. There should be a development of our national forests and

other public reservations designed to utilize land for public welfare.

b. There is a great variation in the economic situation of land; that is, its situation with relation to cities, roads, and economic institutions.

Problem: What should be our national policy with relation to the development and settlement of our agricultural lands?

Related Generalizations:

1. Agriculture has suffered from a haphazard development of land.
2. It is desirable that a better co-ordination be effected among the various national and state agencies concerned with the administration of public lands in order that there may be unified action in bringing about the most economical use of our land resources.
3. In our national reclamation policy it seems highly desirable to weigh the advantages of local or regional development against the disadvantages of promoting excessive agricultural expansion.
   a. Broad considerations of economic, social, and military policy may justify reclamation as a means of increasing the land areas, but the United States is not in such a position.
4. The national government has been operating under land laws that are suitable to a self-sufficing agriculture but are out of place in a commercial agriculture.
   a. Land legislation that expands the agricultural area may at times be justified, but at other times it is of great importance to provide the means of taking land out of agricultural use and helping the people who are stranded on submarginal land to move to regions and occupations which will provide the opportunity to make a decent living.
   b. Political units should consider the advisability of acquiring privately owned land which has been abandoned, or which is unsuitable for farming, and using it for forestry and recreation.
   c. There should be a system of grazing control for the public domain outside of the national forests, comparable with and unified with that now in force in the national forests.
   d. Public ownership will always play a large role in forestry.
5. We should endeavor to prevent the numerous and tragic mistakes in attempting to occupy and develop new farms on privately owned land that has been depleted or to re-occupy areas that have been abandoned.
   a. All agencies engaged in land selling might be placed under public supervision to prevent the misdirection of settlement, and the bringing of submarginal land into agricultural use.

b. The state might seek to promote higher standards by providing a service agency to assist private land settlement agencies to investigate the possibilities of successful settlements and by supplying information to prospective settlers.

c. The state might establish experimental, planned agricultural colonies which would take land out of the unprofitable production of agricultural surpluses.

6. Land settlement is an economic function that is vested with a large measure of public interest.

7. Much work needs to be done by states in the way of establishing plans for settling land on a profitable basis to the settler and to the state as a whole.

   a. At the present time the courts can give comparatively little protection against the exploitation of prospective settlers.

   b. Any land settlement which involves the incurring of capital charges necessarily calls for profitable farming to enable the individual to meet such charges.

   c. The prospect of getting good farm land for nothing may be dismissed as a delusion.

8. Our past agrarian policies have been responsible for a continuous over-expansion of agricultural lands.

   a. The policy of land settlement in this country has been to divide the land among as many people as possible.

   b. In general, since the formation of our government, there have been few restrictions on the use of land for agriculture.

   c. The over-expansion of agriculture and the bringing of marginal land into use are not so much due to the natural land hunger of the people as to the stimulation of various interested agencies.

   d. The liberal disposal of the great body of semi-arid land in the western half of the United States has added to our land troubles.

   e. One of the most striking features of the war development of agriculture was its spread into comparatively low-grade lands with fairly extensive methods of use.

   f. Most of the expansion of our agricultural lands takes place in four directions: Drainage, irrigation, clearing forests, and dry land farming.

Problem: What disposition should be made of the submarginal agricultural land?

Related Generalizations:

1. We have in cultivation a vast area of submarginal land at normal exchange ratios.

2. Submarginal land, from the social point of view, is land from which the average farmer cannot realize a satisfactory standard of living.
3. If submarginal lands were withdrawn from production by private interests under favorable legislation, or placed under the control of state or national agencies to be used for the purposes for which they are best suited, at the same time conserving the soil, a great step forward would be taken.
   a. Favorable taxation and credit policies would tend to promote the use of inferior lands now in farms, or likely to be thrown into farms, for the growing of forests and pastures.
4. Where farmers are living on submarginal land the government should facilitate the abandonment of such farms.
5. If regions are found which will not support a satisfactory standard of living, the public should know about this and the cause be definitely ascertained.
6. Land will be utilized down to and often below the margin as prices begin to move upward.
   a. The submarginal producers of agriculture have greater ability to survive than those in most other industries.
7. It is conceivable that the time may come when it will be highly desirable to bring land that is now submarginal into agricultural production.

Problem: How can we maintain the productive powers of our agricultural lands?

Related Generalizations:

1. As a nation we possess extensive land resources, yet in no other nation is the depletion of soil resources taking place so rapidly.
   a. The era of land abundance has gone.
   b. It is seldom that one generation suffers the full penalty of soil loss; the top soil vanishes too slowly to make its conservation a constant and immediate aim in cultivation.
2. All the people share with the farmers the responsibility of maintaining the producing power of the land.
   a. A generous soil makes a good living possible; a niggardly soil breeds poverty.
3. Our present land situation has been aggravated by the rapid creation of sovereign states, each a political unit burdened with the task of developing its resources and with the responsibility of providing for the upkeep of its government, its roads, its schools, and other expenses of a modern state.
   a. It has been necessary to improve our farm land faster than we

have extended it, and to utilize it for crops faster than we have improved it.

4. The available land supply of a country usually increases less rapidly than the population, so that it becomes necessary to resort to land which is either less fertile, less favorably situated, or more difficult to bring under cultivation.
   a. The tendency to utilize the best lands more extensively and to put the poorer lands into a still poorer utilization is evident in periods of boom and good prices.
   b. Farmers are often able to go on, especially during depressions, through sacrifice of their capital assets and through sacrifice of the soil resources.
5. If fertility of soil is to be maintained, most farming must always be, more or less, a developing operation extending over a term of years, and not simply a cash crop proposition that is measurable by returns per year.
6. Diversification and rotation of crops are fundamental to the maintenance of the productive power of the soil.
   a. Too large a proportion of land in clean cultivated crops is disastrous to soil fertility.
7. Nitrogen, phosphorus, potassium, and all other mineral elements of soil fertility lost from the soil can be replaced by the use of fertilizers, but the humus of the surface soil, the crumb-like structure of the top layer, its water-holding capacity, bacterial content, and all the other factors that make it more fertile than the subsoil, can be replaced only very slowly; indeed, it practically never can be restored.
8. The significant loss in soil fertility from the standpoint of the immediate future is the loss of the top soil by erosion.
   a. Erosion is contributing notably to the development of submarginal areas, and an adequate program for reducing erosion will help to lessen this development.
   b. There should be an elaborated national policy in keeping with which land subject to erosion should be placed under forest or grass cover.
9. Land conservation may be promoted through various forms of government subsidies.

Problem: What causes our farm lands to change in value?

Related Generalizations:

1. The value of farm land increases or decreases in two ways: From the way agriculture is practiced upon it; and also from the play of purely economic causes.

2. In the long run, land values are capitalized net income.
3. In our country the selling value of land has been based more upon hope and speculation than the income that the land yields.
4. Overcapitalization of land values, in view of existing money incomes, has characterized much of our farm land during boom times.
5. Land values change slowly and always lag behind changes which increase or decrease the returns from the land.
6. Land values and the interest charge increase as the value of the dollar falls and as rising prices give prospect of increasing returns to agricultural investment.
7. In a period of declining prices land values stand still, or even decline, and owners' equities in mortgaged land increase very little. But in a period of rising prices, land values may rise 8 per cent or more a year, and owners' equities increase tremendously in consequence.

CENTERS FOR UNITS OF STUDY

B. Production

Problem: How can we balance our agricultural production?

Related Generalizations:

1. More of the proposals for fundamental agricultural betterment have related to production than to any other phase of agriculture.
   a. We know much about how to produce, but far too little about what to produce and the quantity to produce.
2. Land, livestock, labor, equipment, and managerial ability are the agencies of agricultural production.
   a. All agencies of agricultural production have this common characteristic—they vary in their capacity and in their efficiency.
   b. Agricultural production is a method of living, an occupation, a means of employing labor, and an investment; such is the meaning of agricultural production to different groups in our population.
3. Agricultural production seems to be inevitably unstable.
   a. The unorganized competition among the numerous and widely scattered individual producers has been constantly tending to create maladjustments in production.
4. Production of farm products in excess of effective demand has been a recurring cause of low returns in farming.
   a. The farmer's way of getting out of debt has been to work harder and produce more.

   b. The farmer's intention is to increase his production so long as he hopes it will be profitable to do so.
   c. The potentiality of production is increasing faster than population.
   d. The price received for the production of the preceding season has been the dominant factor in the change in production in any given year.

5. We are entering a stage of effective and widespread application of scientific methods applied to the production of agricultural products as a whole.
   a. Despite the depletion in land resources, the trend of agricultural production of the nation as a whole has been rapidly upward, the losses in soil fertility having been more than balanced by advances in agricultural technique.
   b. Improvements in agricultural technique, as well as changes in prices of farm products, are constantly shifting the margin of crop production.
   c. The enormous increase in agricultural production during the war demonstrated that when conditions justify, American farmers can greatly increase the production of most farm products.
   d. Periods of depression give stimulus to invention and bring forth new and more effective means and methods on the physical production side.
6. It is now common practice for the farmer to produce mainly for the market and depend upon income from sales as a means of supplying family wants and of meeting other expenses.
   a. The surplus grows out of the remarkable transition from the old hand-craft, self-sufficient type of agriculture to a modern commercial system.
   b. A surplus is inevitable and desirable if we would have a nation self-sufficient in the material essentials of life.
   c. No program of limitation should eliminate the possibility of a surplus.
   d. Agricultural surpluses may be classified as day-to-day, short-time, seasonal, crop-year, and long-time, or those persisting over a period of years.
7. Production on an ideal basis should be orderly; that is, adjusted to consumer demand as to both quantity and quality of product.
   a. Efforts to improve production must be supplemented by improvement in marketing.
   b. Production limited to the actual needs of the current year can not be carried on with any assurance of a sufficient supply for future years.

c. Men have been producing food for some thousands of years, yet in all that time have not accumulated a surplus sufficient to feed the world for a single year.

d. Complete adjustment is impossible at any given time, due to inability to control annual fluctuations.

e. Adjustments of production would be materially less complex if each farmer made exactly the adjustments in crop and livestock production needed for him to play his part in adjusting the agriculture of the nation to the changed conditions.

f. Although in the long run the farmer tends to adjust his production more nearly to demand, yet he has produced first and found his market subject to every fortuitous circumstance of production and distribution afterwards.

;. We should try to find the way to allow the farmer the full satisfaction of producing bountiful crops, increasing them every year if he can do so, without penalizing him for the effort.

a. The individual farmer has generally been the final determinant in matters of acreage.

b. A popular reason for curtailing production has been our inability to make foreign and domestic consumption keep pace with it.

). Some type of regulated production seems to provide the best visible solution to our problem of a large surplus.

a. The practicability of controlling the agricultural output varies tremendously with the product.

b. Control of production at the lower margin of production would have a tendency to restore prosperity to the good farmer on good land.

c. When all growers alike are made to limit output there is a discrimination against the individually efficient farmer.

d. Under past conditions reduced agricultural production, insofar as it was controlled consciously, was mainly the result of a prolonged period of low prices.

e. Regulation of farm production, to be effective, must begin before crops are planted and livestock is bred.

f. Reduction of acreage in production is difficult unless the released land is used for some other purpose.

g. The policy of crop reduction is sound as long as any farm crop is being produced at a loss.

h. One way to help avoid overproduction in agriculture is to keep out of cultivation land that is too poor for remunerative farming.

i. Even though the total acreage in crops is kept under control, the shifting of acreage as between various branches of production under the influence of price changes may upset the equilibrium of agricultural income.

10. The only cures for persistent overproduction seem to lie in the restriction of output plus expansion of demand as may be possible.
11. Voluntary restriction of any part of the group has been the occasion for expansion by the remainder, and was thus merely a device for shifting income out of the hands of those trying to restrict production into the hands of others.
    a. The production of livestock and livestock products has been more nearly under the voluntary control of farmers than the production of crops.
12. The agency to limit the production of any industry so intensely competitive as agriculture must come largely from the government, with an intelligent land policy as the basis for limitation.
    a. Effort toward balancing the output of the major crops and of the major animal products as well seems doomed to failure unless sponsored by national or state government agencies.
13. From the viewpoint of national agriculture there is a broad educational campaign which should be continuously conducted, looking toward the molding of the volume of the agricultural output.
14. Ultimately the law of supply and demand will effect a readjustment of production.

Problem: To what markets shall we adjust our agricultural production?

Related Generalizations:

1. We should make production adjustments for domestic needs first of all and for exports second.
    a. There has been a declining domestic demand for agricultural products.
    b. Our nation should produce those agricultural products which in the long run will add most to the well-being of the people.
2. Looking back over the market prices of nearly any season, farmers can see how their incomes might have been greatly increased if they had sold their products a little sooner or a little later.
3. The farmer has always had a market in which his staple crops could be sold.
4. The development of modern systems of transportation, the growth of great industrial cities, and the expansion of specialized commercial agriculture has brought with it a complex middleman system of marketing.
5. The agricultural industry largely retains its individualistic methods

in selling practices, although there has been a great gain in co-operative selling.

6. The demand for agricultural products is relatively inelastic or constant with a given population.
7. There is great possibility of discoveries which will find new uses for agricultural raw materials.

CENTERS FOR UNITS OF STUDY

c. Distribution

Problem: What are some of the most notable defects of our distribution of agricultural products?

Related Generalizations:

1. There is an enormous cost in the distribution of products.
2. For the most part, our marketing system consists of a very large number of small units operating independently of one another, lacking in integration.
3. Our marketing system essentially remains in the hands of those interested in the portion they extract for themselves along the line of distribution.
   a. In the main, the functions of the middleman are little understood by the consumers.
4. Short-time surpluses and shortages exist with respect to many products in different parts of our country at any given time.
   a. No matter how efficient the marketing system may be, its benefits might be completely nullified, as far as the producers are concerned, by an overproduction of the commodity.
5. The location of present rail and water terminals of the wholesale markets for farm produce was in most instances determined by conditions which prevailed perhaps a century ago.
6. The gathering for shipment from many isolated growers is more expensive than the assembling of the product from a region of specialized production.
7. There is a tendency to maintain a flat-price system rather than pay for quality.
8. The farmer has not had full knowledge of the marketing situation and hence could not make wise individual effort in marketing.
9. The attitude of the individualistic farmer has often proved to be contrary not only to profitable marketing but also to production.

Problem: How shall we distribute agricultural products to consumers?

Related Generalizations:

1. Co-operative co-ordinated marketing, and in some cases warehousing, of surplus farm products at times is desirable.

2. Some constructive help to the farmer in marketing his goods at a profit can be found through the development of producers' marketing organizations.
   a. Without group action the farmer will have neither the knowledge nor the power to produce a product of quality and place it where it will command a price commensurate with its quality.
3. The expansion of markets, both at home and abroad, and the development of new uses for agricultural products are among the methods of meeting the surplus through marketing.
4. There should be substituted a number of centralized selling agencies for the principal farm products, if we would bring agriculture into its proper relation with the modern industrial and commercial world.
   a. So long as we have competitive producing units we need not fear monopoly power from co-operative marketing organizations, because monopoly springs from a control of production or consumption.
5. Private, competitive marketing agencies have a useful place in the distribution of our agricultural products.
6. Grading, standardization, inspection, and certification of products will help in marketing.

Problem: What are the principal effects of modern transportation on the agricultural industry?

Related Generalizations:

1. Any change in the methods, facilities, or cost of carrying farm commodities produces very definite influences upon the industry.
   a. The nature of the agricultural industry is such that rapid adjustment to changed transportation situations is not made.
   b. The farmer is peculiarly affected by transportation costs in that he cannot fix the place of production at the point most advantageous to marketing.
2. Railways, highways, and waterways have put a given specialized producing area of our agricultural plant into competition with other like areas.
3. The interdependence of domestic and foreign marketing rests upon the modern system of transportation and communication.
   a. Transportation facilities are of the greatest importance in determining the ability to market and where products may be marketed.
4. Transportation charges in the long run are included in the costs which the consumer must pay if the needed supply is to be forthcoming.

  a. Transportation costs account for much of the spread between producer and consumer.
  b. Transportation rates in this country have favored large established centers at the expense of small centers.
5. Freight rates should be determined from the point of accomplishing the ends desired, which are largely social in character, and should be evaluated from the standpoint of long-run economy.
  a. The freight rate structure, which is the outgrowth of many factors involved in the industrial agricultural development of the country, is so complex and interdependent in its parts that it is difficult to readjust quickly to changing conditions.
6. The outstanding development in the field of transportation is the greatly increased use of the automobile, which brings with it the need for suitable highways.
7. Inland waterways, expected to bring about a reduction in transportation charges on staple farm products, are subjects of study now as to their feasibility and economy.
  a. The farmer has long been interested in the development of inland waterways as a means of reducing and equalizing transportation costs.
  b. The drift in transportation has been away from the use of waterways to other speedier forms of transportation.

CENTERS FOR UNITS OF STUDY

D. Prices

Problem: What importance is attached to the price of agricultural products?

Related Generalizations:

1. The most popular criterion of agricultural prosperity is the price level of unit quantities of farm products.
  a. The farmer sells all his labor in the form of his products and the prices of his products are, therefore, a sensitive gauge to his status in the general markets.
2. The farmer is not concerned directly with the wholesale and retail prices in cities, but with the prices that he gets for his products, and with their purchasing power.
3. A significant thing in the position of agriculture is not only what its products are worth in terms of dollars, but what these dollars are worth in terms of the goods and services necessary to produce these products.
  a. A change in the purchasing power of the dollar makes fundamental changes in price relationships in different parts of the

channels of trade, which are entirely separate from the question of supply and demand.

Problem: What causes increase or decrease in agricultural prices?

Related Generalizations:

1. The price which the American farmer receives for his products is determined largely by the supply and demand for them.
   a. When individual farmers put their product on the market at the same time they render it a homogeneous supply and, therefore, in technical terms, the marginal price—the price of the final unit added to the supply—is the price for which the whole crop sells.
2. The size of the crop has more influence on price than flow to market.
3. The peculiar characteristic of agricultural prices is their extreme sensitiveness—what appears to be a minor cause produces disproportionate effects upon them.
   a. It is characteristic of an inelastic demand that a little surplus greatly depresses price and a little deficit greatly advances price.
4. Efficiency increases production, and, as production increases faster than the population, prices tend to fall.
   a. Maladjustment of production and consumption and faulty methods of marketing are often important factors in price declines.
5. Indirect competition of industrial and agricultural substitutes is playing an increasingly important role in the pricing process.
6. Other things being equal, the rise of urban wage levels tends to increase the demand for manufactured products faster than that for agricultural products and so raises the price levels relatively.
7. The further the product is from the consumer and the more hands it must pass through, the greater the discrepancy between farm prices and retail prices whenever there is a change in the price level.
   a. The spread between prices on the farm and prices at retail is much greater in the United States than in European countries, because of higher wages, higher transportation costs, greater loss from spoilage, and greater service demanded by the American consumer.

Problem: What are the effects of increased or decreased prices, of inflation and deflation, on agriculture?

Related Generalizations:

1. Expansion and contraction of currency produce fluctuations in

prices of articles bought and sold by farmers and become a subject of controversy.

2. Primary commodities, such as farm products, rise more quickly when the price level is advancing and they fall more quickly when it has a downward trend.
3. When there is a rise in the price of farm products the consumer feels it at once, but a decline takes a much longer time to be passed on to him.
4. Increased prices tend to decrease consumption and increase production.
   a. When the farm price is cut in half, the farmer does not get the benefit of the increased consumption that would follow a halving of the retail price.
5. Supply and demand remain in adjustment at the consumers' price, but when inflation or deflation occur, the farm prices are often more influenced by the change in the general price level than by the changes in supply and demand.
6. Inflation without any change in supply or demand automatically stimulates agriculture, and deflation without any change in supply or demand depresses agriculture.
7. The lag of wages, and the lag in taxes and interest, are the major causes for the innumerable maladjustments that result from inflation and deflation.
8. With the slow investment turnover in agriculture and the small cash accumulation characteristic of the industry, it is difficult for the farmer to adjust himself to inflation and deflation.
9. Since all the purchases of farm property during a period of inflation are at prices of the day, those who contract debts to be paid after deflation has occurred must pay much more than they anticipated.
10. Index numbers of prices paid to producers of farm products in selected states indicate that in a general way incomes in the different regions have followed the course of prices of farm products.

Problem: What effect does the world price of agricultural products have on the domestic price?

Related Generalizations:

1. Whenever we export agricultural products, our prices participate in the making of world prices, and world prices react on the prices in each exporting country.
   a. Irrespective of whether there be an export surplus or a deficiency, and irrespective of a balance between domestic supply

and demand, American wholesale prices have remained more or less closely bound to the world markets.

b. The degree and quickness of response to foreign price influences vary somewhat with the portability and keeping quality of the farm products and with the adequacy of the marketing system.

2. Prices at which the world will buy, accompanied by lowered costs, are an essential to the welfare of the farming industry.

   a. The price at which the exportable surplus of our agricultural products can be sold in a large measure governs the price of the whole.

3. In the case of commodities grown all over the world and subject to world market conditions, the price received per unit of the crop has little relation to domestic yields.

4. Shifting from an export to a domestic market basis may make a material change in price relations for a producing area.

Problem: How can we more nearly stabilize the price levels for agricultural products?

Related Generalizations:

1. Wide fluctuation in the price of a commodity usually results in a loss to the producer and the consumer alike and stabilized prices benefit both.

   a. Prices of farm products fluctuate more under our present system of commercialized agriculture than formerly.

2. The farmer is a seller of a basic commodity and prices of such commodities are apt to fluctuate more violently than the prices of other goods because the supply of farm goods can expand and contract much less rapidly.

   a. Farm prices fluctuate more than wholesale prices, and wholesale prices fluctuate more than retail prices.

   b. If a crop product fluctuates widely in price there is a tendency for the retail price in cities to remain high enough to make the merchandising of the crop profitable when the wholesale price of the product is high.

3. Prices of different commodities are usually fluctuating, but generally they are fluctuating about the same base level.

4. The success of an effort to stabilize farm income by reducing fluctuations in price from year to year depends in a large measure on the stability of the general price level.

   a. Domestic stabilization is hardly feasible without stabilization of prices abroad.

5. The price of a commodity can be stabilized only at a point near the cost of production.

6. As long as an individual farmer sells his small quantities, interposing his individual selling power against the organized buying power of other industries, he will continue to be the main sufferer from price fluctuations.
7. Stabilized prices may be more nearly sustained when producers of given commodities agree to regulate production.

CENTERS FOR UNITS OF STUDY

E. Credit and Income

Problem: What are the credit needs of agriculture?

Related Generalizations:

1. The farmers' credit needs are short-time, long-time, and intermediate.
2. Agricultural credit is of such a nature that ordinary credit agencies do not provide it properly.
   a. The need for a better system of farm credit, both mortgage and personal, was felt for many decades, and resulted in national farm credit.
   b. The bulk of short-term credit used by farmers is supplied by rural banks and merchants.
   c. Commercial banks have probably been the weakest part of our agricultural credit system.
   d. The close dependence of agriculture on nature makes risk an especially important element in provision of credit for it.
   e. The provision of credit for agriculture is greatly affected by the close relations between the family living and the farm business.
3. There is a need of adequate facilities for financing the requirements of operating capital.
   a. In addition to mortgage credit, which is most commonly used for the purchase of land, most farmers need to borrow on personal credit to secure sufficient working capital and make needed improvements.
4. It is a common practice in the United States for farmers to borrow money to invest in land.
   a. The effect of a good credit system is to encourage land ownership on the part of farmers.
   b. Mortgages on farms are generally used as a means of making the transition from tenancy to landownership, and are frequently an evidence of growing prosperity.
   c. Mortgage obligations should be scaled to the actual earning value which agricultural lands will have during the working

lifetime of the present generation as nearly as this can be estimated.

d. Farm mortgage debts are of long duration, requiring often a lifetime for liquidation.

e. Variations in commodity prices often make it necessary to take a far larger proportion of the total crop to meet mortgage obligations than it did when they were incurred.

f. One of the most serious phases of the mortgage situation was the heavy indebtedness incurred by farmers during and after the war.

5. Credit, properly directed, can bring about substantial improvement in the economic status of the individual farmer, raising his standard of living and decreasing his dependence upon credit advances.

a. Under our credit system the borrower's activities are often controlled by the lender.

b. To promote their own welfare, urban interests must think of getting money out to the farmer as well as getting money out of him.

6. With the growing commercialization of agriculture, the use of credit by farmers has greatly increased.

a. The great majority of farm business transactions involve, not the use of currency, but of bank credit.

b. Flexibility of credit and of interest rates are of the greatest importance to the farmer.

Problem: How should agriculture share in the national income?

Related Generalizations:

1. The prosperity of the agricultural industry as an operating proposition depends largely on the purchasing power of its own income in terms of the goods and services which it uses.

a. Income not only determines the amount of money which families can spend on their living, but also the amount of taxes that can be raised for education, roads, and other local improvements.

2. Agriculture thinks that it is not getting its share of the national income in proportion to the labor expended and the capital invested.

a. It is to the establishment of the higher standards of living on which the rural population will insist, if they are to continue to produce agricultural products for the market, that we must look for the real foundation of any movement looking toward a permanently larger share in the national income.

b. Such modifications in our laws and institutions should be brought about as will make possible a distribution of the national income which will enable farmers to live as well as other groups of people who apply equal intelligence, skill, energy, and capital to their productive efforts.

CENTERS FOR UNITS OF STUDY

F. Related General Economic Factors

Problem: What is the general economic pattern in American agriculture?

Related Generalizations:

1. American agriculture consists mainly of individualistic enterprises by families.
2. Modern American agriculture is organized largely according to the scheme of freedom of enterprise individually directed and independent of supervision from any authority higher than its owner.
   a. The history of civilization has indicated that the way to get the most out of land is under private ownership, with the appeal to individual initiative and ambition.
3. There has been, and is now in process, a veritable revolution in agriculture which may prove to be comparable in scope and significance with the industrial revolution.
   a. The definite trend of American agriculture is toward industrialization.
   b. We are tending to modify the traditional individualistic system for agriculture with some collectivistic activities.
   c. The self-contained and self-sufficient agriculture of pioneer days is not going to return.
4. If agriculture is to be readjusted, those changes which come can only be induced by the establishment of right economic incentives.
   a. It is a complacent delusion that the evils under which the farmers suffer are inevitably self-correcting.

Problem: What is the economic relationship between agriculture and our other great national industries?

Related Generalizations:

1. Agriculture furnishes the rest of the industrial system with much of its raw materials and in turn expects the manufactured goods it needs.
   a. Although agriculture has taken second place in our total productive life, it is still a large and often a determining factor,

not only in the general welfare but particularly in the prosperity of the urban classes.

2. Agriculture and other industries have important mutual relationships and influences.
   a. The hard and fast line between agriculture and other industries is tending to disappear.
   b. If any occupational division of the population, say agriculture, is deficient in buying power, other divisions cannot be indifferent to that fact.
   c. Difficulties for agriculture and the other kinds of industry arise from the fact of the shifting significance of each in relation to the other.
3. There must be a co-ordination of agriculture and other industries in the interests of a balanced economy.
   a. Organized business is co-operating with agriculture on a much larger scale than either the average farmer or the average business man fully realizes.
   b. If the reality of industrial life and the reality of agricultural life could meet face to face, the understanding of both would be illuminated.

Problem: What is the influence of costs in the agricultural industry?

Related Generalizations:

1. Agriculture is changing from high cost to low cost of production.
   a. One of the most effective means of reducing production costs is to increase the yield of product.
   b. One way to lower cost in farming is for each agricultural region to produce what its natural conditions best fit it to produce.
   c. One essential point of commercialized agriculture is a grouping in commodity production so that costs may be decreased.
   d. The individual farmer can cut his costs through the efficient operation and management of his farm.
2. The enormous reduction in labor costs per unit of production which has taken place through the application of improved farm machinery is one of the significant changes in American agriculture.
   a. The value of the labor done by the farmer and his family and the interest on the farm investment constitute a large part of the cost of production on the farm.
3. In the simpler age, cost of production did not concern the farmer much. . . . Today all is changed.
   a. Farming is like any other business in that only by careful

analysis of costs can the influence of any one factor of management upon the success or failure of the enterprise be determined.

4. When agricultural costs are viewed as a whole, it appears that the primary factor in the discrepancy between the level of costs and that of income is the high overhead cost per unit of product due to the high level of taxes and debts, combined with the relatively low yield per acre.
5. There is great inelasticity of agricultural costs in relation to prices received for farm products.
   a. In farming, as in other industries, the cost of production tends to approach the selling price, and the selling price to approach the cost of production, for the marginal producer.
   b. Cost of production figures of agricultural commodities are always difficult to determine because of the many intangibles in farming.

Problem: How has the factor efficiency influenced the agricultural industry?

Related Generalizations:

1. In the long run agriculture will benefit from increases in efficiency in its own processes; but over short periods and under particular conditions it may benefit little, if any.
   a. As industrial efficiency is the foundation of the high scale of living of American labor, so agricultural efficiency may be expected in the long run to increase the scale of living in agriculture.
   b. In the long run the effect of increased agricultural efficiency is to increase the city population relative to the farm population.
   c. As inefficient producers are eliminated but slowly from agriculture, improvements in production often have for a long time the result merely of increasing the output beyond market demands and of thereby driving prices lower than they were before.
2. Production is easily expansible with no reasonable limit and the competition for money profits is intense, tending always to limit profits for the producers of high efficiency and to destroy profits altogether for the producers of low efficiency.
   a. Efficient farmers prosper beyond the inefficient.
3. A great increase in efficiency in producing food will result in enabling a smaller number than formerly to meet the present demand.
4. Governmental agencies are interested in increasing efficiency in

farming so as to reduce costs rather than increase volume of production.

a. The efficiency of agriculture today is such that the country can readily produce supplies beyond the consuming power of existing markets, if sufficient inducements are brought to bear to stimulate a maximum output.

Problem: How has machine development affected the agricultural industry?

Related Generalizations:

1. Power machinery, making it possible for one man to have control over large acreages, is a relatively new development in agriculture.
   a. With the exception of transportation, American agriculture has more power available per man than any other industry.
2. Mechanized methods could meet the present demand for food products even though thousands of small farmers dropped out.
   a. Engineering developments in connection with power farm equipment and farm organization have greatly reduced labor requirements and unit costs.
3. Mechanization tends to make farmers more dependent on the whole financial and commercial structure.
   a. The mechanization of agriculture reduces the farm market for many farm products because of the reduction in numbers of farm work animals.
4. The possibilities of using machinery have much to do with the income the individual can earn in the various types of farming.
5. In the utilization of machinery, co-operation is becoming more common.

Problem: How is labor employed in the agricultural industry?

Related Generalizations:

1. Farming in the United States has been largely a story of efforts to economize labor rather than land.
   a. In production for labor expended, the supremacy of the American farmer is probably greater today than ever before.
   b. Much of our successful farming rests on the unpaid labor of women and children.
   c. Child labor still runs high in rural districts.
2. Although about three-fourths of the labor of agriculture is normally done by the farmer and his family, the wages of hired labor are an important factor in the economic position of agriculture.
   a. It is doubtful whether farm wages ever have been high enough

in the United States to warrant any man in taking up farm labor as a life occupation.

b. Wages of farm labor are a compromise between the prices of farm products and the wages in other industries.

c. High urban wages work to the disadvantage of the farmers in that farmers are not able to compete with other branches of industry for labor.

d. Farm wages respond to prices much more quickly than do wages of industrial workers.

e. A period when food is very cheap compared with wages results in drastic steps in the saving of labor on farms.

3. Unemployment, far from being a strictly urban phenomenon, occurs in agriculture.
4. Obtaining an adequate supply of labor is most difficult in highly specialized sections and where the demand is intermittent.
   a. Labor is usually absolutely limited in pioneer farming.
   b. There never has been such a supply of farm labor but that still more of it would have been used if it were cheap enough.
5. A significant proportion of the immigration not now restricted by quota tends to enter the agricultural industry rather than others.
   a. It is undesirable to advance the agricultural industry by the importation of cheap labor, either of a temporary or a permanent character, that is not easily assimilated.

Problem: What type of farming shall we have?

Related Generalizations:

1. Types of farming that prevail in a given region are determined very largely by forces which are beyond control of the individual.
2. The practice of utilizing much land, that is, extensive rather than intensive cultivation, prevails in American agriculture.
3. American agriculture holds a place for both specialized and diversified farming.
   a. Specialized production by soil areas and climatic zones should be encouraged wherever such specialization would be in the national interest. On the other hand, diversification should be urged where specialized or single crop farming is detrimental to the land and to the economic and social interests of our farm population.
   b. Small areas and great diversity may be desirable in limited localities where production of home products is essential to a good food supply for the farm family, but commercial production demands specialization and concentration.

4. Diversified farming lends itself best to that form of rural life which sociologists and historians wish to retain as a part of the American heritage.
   a. Diversification not merely reduces the acreage devoted to the principal crop of the region, but it also spreads the farmer's risk so that failure of a given crop from any cause is not so disastrous.
5. If the farm is to be brought parallel with factories, it will have to be an all-year plant, which means intelligent diversification.
6. Increased specialization in agriculture is taking place at a phenomenal rate; this in spite of urban advice to the contrary.
   a. As we specialize more we increase the hazards of agriculture.

Problem: What kind of farm organization should we have in America?

Related Generalizations:

1. There are but two types of successful farming in the United States today under present economic conditions, namely, the farm on which the farmer and his family do nearly all the work and there is no payroll, and the larger corporate type farm.
   a. American agriculture is tending to smaller as well as to larger farms.
2. The trend toward large-scale farming is not necessarily a trend toward corporation farming.
   a. One development in farm organization which has received much attention is what is commonly called "chain farming"; that is, the operating of a string of farms in one locality not too far removed from each other, under the close supervision of a landlord or general manager, with tenants or hired operators on each.
3. Large units may operate successfully in certain areas and with certain products, but the great bulk of agricultural production will take place on family-sized farms.
   a. The prospect of factory methods being used in agriculture has social implications comparable to those of the industrial revolution.
   b. The large-scale farm is opposed, not because it is thought to be unprofitable, but because it threatens to remove the last great bulwark of American individualism.
   c. In the nature of things, the small, isolated farm unit is often not in a position to realize most of the opportunities of large-scale production.
4. Nothing like the rapid consolidation of enterprises that often takes

place in the business world is profitable in agriculture; and it may be taken for granted that the development of large-scale agriculture, at whatever rate it may proceed, will penalize the very small man less severely than similar tendencies usually do in other occupations.

5. Subsistence farming becomes imperative to the average American farmer as a means of employing unoccupied time that cannot otherwise be remuneratively disposed of.
6. The true industrialization of the farm will come with a shift in the point of view—from the work in man terms to the work in power terms.
   a. An industrialized farm system would aggravate the present industrial evil of unemployment.

CHAPTER IV

INTERPRETATIONS AND USES OF THE MATERIALS

THE principal contributions of this study are to be found in Chapters II and III. This fourth chapter deals with some possible interpretations and general uses of the materials presented, while the fifth chapter gives a more concrete application of the materials in the form of a course of study unit.

THE TECHNIQUE USED

The technique explained in Chapter II is one of several that teachers may use in the development of materials for a study of our contemporary problems. Though probably few teachers would wish to go into so much detail or go into so extensive a piece of work as this study, many could use this particular technique in the development of certain curriculum materials for social studies. It involves the use of many sources, but good teachers are not now relying on single textbooks, single pamphlets, or single observations for their consideration of contemporary problems. The element of complexity alone would lead to a wide sampling of data and opinions. So the technique described earlier in this study may be put into practice on as limited or as extended a scale as teachers may wish. But the materials (rather than the technique) developed in Chapter III may be more widely used by teachers and other curriculum workers.

MATERIALS FOR SECONDARY SCHOOL TEACHING

The materials are designed primarily for the use of secondary teachers in our American schools. They are probably not complete or extensive enough for use in college courses; nor would they be suitable for use in elementary schools because of the difficulty of the language used and the complexity of the problems noted. During recent years there has been a growing interest in the consideration of contemporary problems in our secondary schools. Some schools have organized courses for the study of modern problems, some have included the study of such problems in their social studies courses, and others have followed different procedures. Several publications have been issued for the assistance of teachers.[1]

"PROBLEMS" IN THE CURRICULUM

In Chapter III special forms of subject matter are used. First, there are certain stated *problems* of the agricultural industry, and second, there are related *generalizations* which have some bearing on solutions of the problems. One may utilize such stated problems in different ways. There is a danger that teachers may incorporate them, as suggested in Chapter III, *in toto* in a course of study. They are stated for another purpose; that is, they should be used more as guides by teachers who may be interested in a study of agricultural problems. Why? The problems as stated may not be genuine for pupils; and, as Dewey[2] says:

> . . . it is indispensable to discriminate between genuine and simulated or mock problems. The following questions may aid in making such discrimination:
>
> a) Is there anything *but* a problem? Does the question naturally suggest itself within some situation of personal experience? Or is it an aloof thing, a problem only for the purposes of conveying instruction in some school topic? Is it the sort of trying that would arouse observation and engage experimentation outside of school?
>
> b) Is it the pupil's own problem, or is it the teacher's or textbook's problem, made a problem for the pupil only, because he cannot get the required mark or be promoted or win the teacher's approval, unless he deals with it?

There is always the possibility that teachers may first adopt the problem form of subject matter and then assume genuineness for the learner. These problems stated in Chapter III may simply need a change in language so as to become real for the learner,

[1] For example, Herbert B. Bruner, *Some Suggestions for the Study of Modern Problems. A Bulletin for Teachers.* Bureau of Publications, Teachers College, Columbia University, New York. 1934.

[2] John Dewey, *Democracy and Education,* p. 182. 1915.

and again there may be a need for a more fundamental change. The real problem for purposes of curriculum making is one that is recognized by the learner as being of importance to him; through its study and solution he will have realized the answer to a genuine felt need. The activities carried on in solving the problem, then, have an accompanying element of purpose, moving the learner toward an end he believes to be worth while.

Problems which are not at first real or genuine to the pupil may be made so by the skillful teacher through appeal to certain aspects which may be interesting at the time. For example, pupils may not think of the distribution of farm products as a genuine problem to them, but a derived interest may be aroused and a genuineness created through field trips to marketing centers. Counts[3] declares:

> Learning is prosecuted most effectively when the individual identifies himself most completely with the thing to be learned. Only under these conditions is there neither dispersion of attention, nor dissipation of energy. Yet, that the purposes of education can be achieved wholly through interest of this type is a fatuous deception. A fair measure of the work of the world must be performed under the conditions of derived interest.

WHAT PROBLEMS SHOULD BE CONSIDERED?

The problems stated in Chapter III have been organized into three divisions, namely, Political, Social, and Economic. Within each division there are indicated several centers for ***units of study.**** Since units of various types seem to hold the educational stage now, there are suggested certain centers around which units may be developed. These centers are logical subdivisions of the subject materials and are points of reference for the problems which are listed under them. A given problem, as stated, may become the title for an individual unit of study, or the teacher may combine the statements of several problems and use this new statement as the title for a unit. Still others may wish to

[3] George S. Counts, "Some Notes on the Foundation of Curriculum Making," *Twenty-Sixth Yearbook of the National Society for the Study of Education*, Part II, p. 79. 1927.

* See Chapter V for a development of the "unit of study."

use the "Centers for Units of Study" as the basis for titles. It would probably be undesirable for any secondary school class to consider all the problems stated in this study, because, in the first place, there is doubtless insufficient time for such consideration, and, again, not all these problems would fit the needs or interests of the group. However, there are certain of these curriculum areas (and their related problems) of such importance that they should receive some consideration in all public American secondary schools, whether urban or rural. These curriculum areas are as follows (in the order of their appearance in Chapter III): The National Government, Rural Population, Land, Production, Distribution, and Credit and Income.

Naturally, these centers or areas are not of equal importance to different groups in our population, and it would be virtually impossible to judge which is of greatest importance. *Land* may be of most controversy and interest in certain sections of our country, as in the southwestern states; *distribution* may be equally controversial and interesting in other sections, as in New England. If a teacher will start a consideration of one area of great immediate interest, it will be very difficult not to filter into the other areas. There is little doubt that some one of the six areas selected would furnish an excellent point of reference.

In order to gain a measure of appreciation of these six important curriculum areas, brief accounts of possible subject materials for each area are included here.

1. *The National Government* (*Its Relation to Agriculture*)

Not even in agriculture are our American problems wholly economic, wholly social, or wholly political (as the main divisions used in Chapter III might lead one to conclude). Reference to the problems of the National Government will show that there are definite economic implications as well as political implications considered in the list (even though they appear under the general division of political problems). Chapter III shows for the National Government these problems:

What should be the general function of government with respect to the agricultural industry? What should be the tariff policy in relation to agriculture? To what extent should our public agencies enter into the fixing of prices on agricultural products? To what extent should our public agencies enter into the distribution or marketing of agricultural products? To what extent should our public agencies enter into the control of production?

All these problems are permeated with social-economic-political issues and views. They are all highly controversial and hence subjects of debate between economists and politicians, between conservatives and radicals, between farm organizations and business organizations. It is no wonder that the novice, when he begins to read in that field, is bewildered with proffered solutions which are diametrically opposed.

The tariff has been an interesting political-economic measure of vital importance to agriculture since the middle of the nineteenth century. Calhoun stirred the nation with his attacks then, and once again the cotton planter is crying out against what he thinks is an obstacle to international trade. Through tariff barriers set up by our National Government and by the central governments of other countries, much of our foreign agricultural trade has been cut off. Business interests have dominated the tariff policy for a long time. They have protected their own growing industries and have let agriculture suffer unduly. Such a short-time view seems incredible to the farmer. Much of it is based on a very inadequate understanding of the problems of agriculture by business leaders and by pressure groups which often force legislation in Washington. One has only to read a portion of their literature on farm problems to get this picture; but a far more important illustration is to be found in the legislation which they have sponsored.

The Jeffersonian idea of independent farmers accustomed to little regulation or control is a vastly different idea from that recently advanced under the AAA. Great changes in the role to be played by our National Government have taken place, even within the past decade. The idea of planning in the interests of farmers and the agricultural industry was suddenly taken over by agencies in our National Government. To these agencies was entrusted, in a large measure, the administration of ideas embodied in laws by Congress. Under the stress of economic hardship one of our greatest bulwarks of individualism (the farmer) submitted to a type of state control and regulation. Farmers have been studying the factor of planning in their industry and farmers are willing to plan, but many farm leaders are reluctant to see these plans made and carried out by a government which they believe to be so greatly dominated by the big financial and business interests. Many believe that farmers as a group have been so ineffective politically in our National Government that it would be unsafe to put such vital matters as control of production in the hands of any one central governmental agency for any length of time. Such proposals, they think, smack of Fascism and not of democracy; they represent the seeds of disintegration of democratic control. On the other hand, many who are equally sincere and positive believe that national planning by governmental agencies is our only hope for the future. A critical study of these and other views is needed in our public secondary schools where many of the problems of vital importance to our national welfare should be considered.

2. *Rural Population*

In order to understand better the problems of the agricultural industry, it is well to gain a better view of the rural mind. Farmers are generally conservative in their social-political-economic views. They are innately individualistic, primarily because of the nature of their occupation and because of the American way of settling our agricultural lands; yet their many co-operative enterprises furnish an interesting contrast to this characteristic. Property rights are of paramount interest to them, and they represent a vital part of our capitalistic system. They are slow to change their ways, partly because of their love of tradition and conservatism, but mainly because of the fact that farming is not only their way of making a living but also their way of living.

Throughout most of the history of our country there has been a migration of population from the farm to the city. There are many reasons for this movement, but among the more significant are these: a much higher birth rate in the country; possibilities for greater social and economic advantages to individuals living in the city; changes in the science of agriculture, making possible the release of great numbers of agricultural workers for work in other fields; and changes in the technology of manufacturing and business.

During the past few years of depression there has been a movement in the opposite direction, from the city to the farm. The movement from city to farm and the decentralization of industry have been advocated by some of our more prominent business leaders, but have met with little support from most farm leaders. Certainly the addition of more people to the ranks of agriculture is in opposition to a major trend in population movement. Our National Government has begun to take part in the movement of population in many ways, but among the more interesting is its Resettlement Division. These movements to the farm bring to the front that problem of just how many workers there should be in the agricultural industry. Insofar as one might consider present demand for agricultural products in America there is probably a surplus of producers, but it is very questionable whether or not there would be a surplus if there were a more favorable distribution of the products.

The standards of living and scales of living for rural people have been the object of much study. One interesting approach to this problem has been through a study of the effect of land ownership and land tenancy. Although tenancy is often a means of acquiring land, and thus useful, it is too often the symbol of poverty of living in all its many aspects. It is because of the latter fact that sociologists generally are deeply concerned. Underlying the existence of tenancy are several other great economic and social problems that are attracting serious study; such problems as farm credit and land utilization.

3. *Land*

For our American people land has great social, political, and economic significance that should be more clearly recognized than is the case now. Recent studies of soil erosion, which is due to water, in the Mississippi Valley have pointed out the necessity for many corrective measures. Where once there were trees and grass to hold the soil, to help prevent floods, there are now denuded areas that present many serious problems of national importance. Studies of wind erosion in our southwestern states show that much land adapted to grazing purposes was plowed under during war times and the native sod destroyed. As a result, there are great dust storms, and land that was formerly useful is becoming barren. Few nations on the earth have been blessed with the abundance of fertile soil that America has had, yet there has probably never been a more profligate user of this inheritance.

Land has been put to many uses in America and will probably be put to more varied uses in the future. There is now under way in our country an attempt to classify all land for particular uses. In most states the agricultural colleges have for many years studied the question of land classification, mainly from the viewpoint of agricultural uses. Although the agricultural industry is our most extensive user of land, it is not the only one. A few of the other uses to which land is put are as follows: as sites for manufacturing plants, for communities, for public recreation grounds such as parks, and for game preserves; as land to be used in the regulation of stream flow; as forest preserves; and as areas to be given over to hydro-electric power development. Great progress has been made in research and in dissemination of information on this question; in this lies hope.

Some confusion in the development of our agricultural lands has been occasioned by conflicting acts of different agencies of our National Government. One bureau has been interested in reclaiming semi-arid areas, thus adding to our national area of productive farm land. At the same time another bureau has been purchasing submarginal land and withdrawing it from the productive area. Much of this conflict is being eliminated through recent National legislation.

The problem of how and when our land should be occupied for agricultural purposes is coming to be of greater importance. Old homestead laws no longer offer a feasible plan because there is relatively little "new" fertile land to which the settler can move.

Most of it has already been settled and is owned or rented. This means that much submarginal land may be occupied unless there is some means of preventing such occupation. Most of our states have followed the path of least resistance and have left this question of occupancy to the individual and to land companies. A few states (notably Wisconsin) have pioneered in an attempt to assist land settlers in the selection of desirable land, and some have even passed laws that would prevent the settlement of undesirable land. Such a procedure has met with opposition from vested land interests; nevertheless the program of regulation is expanding. In the opinion of many it must expand if America is to realize a greater degree of permanency in her agriculture. The occupancy of land and its subsequent management or development are matters of great importance to all in America. Civilization has been a costly user of land, judged by history, and America is in danger of repeating history. America represents one of the newest of the countries occupied extensively by white men, yet in the short period of her history she has seen the abandonment of many once fertile areas. It is well to re-emphasize here that *the problems of land utilization are basic to other agricultural problems. Upon their intelligent solution rests, in a very large measure, the material welfare of our nation.*

4 and 5. *Production and Distribution*

These two curriculum areas are considered together, for it is well to think of them in close relation to each other. Probably more has been written about them recently than about any other (agricultural) curriculum areas that could be named; certainly many legislative proposals affecting them have appeared in the past decade.

The term *production* in agriculture is generally understood to mean the process of raising animals and crops. Land, livestock, equipment, managerial ability, and labor are some of the agencies of production that are commonly recognized. It is through the action and interaction of these various agencies that our foodstuffs are produced and various other ends of production realized. Although in economic theory *distribution* means the apportionment of the gross returns of agriculture among the several agencies of production, yet this is not the meaning most commonly associated with the term by the layman. In deliberate violation of the economic definition, *distribution* is here used to signify the manner in which the products may be transported to the ultimate consumer, and marketed. (This use is particularly true for the generalizations cited in Chapter III.)

The question of what the individual farmer should produce and how much he should produce has been a matter left largely to him for a decision. True, he has had to consider the land, the climate, the market, his equipment, and many other factors. But he has been rarely told that he must produce a certain crop on a specific number of acres, unless he is a tenant. Under the compulsion of share tenancy and of our credit institutions many farmers have had to produce certain crops and animals. This type of control over production is today one of the major types in America. Much has been written in recent years about the advisability of planning production and of controlling production. This writing has been most prevalent in recent years because of the peculiar circumstances of recent agricultural production.

The traditional American view of agricultural production has been greatly changed under the influence of our industrial and technical development. There was a time in our history when the production of individual farms was mainly of those items which would be needed by the family living on the farm. The foodstuffs and the materials for clothing were raised on the farm. Fruits and vegetables were canned for the winter. Products might be exchanged for other necessities of life, but there was little cash income for the farm. Gradually towns and cities developed and local markets for the products arose; railroads spread into the various sections of our land. International trade developed and the demand for such agricultural products as cotton and wheat increased. Then farmers began to think more in terms of what they could produce in abundance and sell profitably. So agriculture changed from a major emphasis on subsistence to a major emphasis on sales. Products in much greater quantity than was needed for mere family requirements were the goal, and these products vied with others for a domestic and a world market. The word *surplus* entered our vocabulary under the impact of the commercialization of agriculture and modern production methods. What to do with and about this surplus was and is a matter of serious concern to many people. There are those who claim that this overproduction is due to a faulty distribution and that there would be no real surplus if each person were supplied the necessities required by a decent scale of living. Since those who consider our production as excessive have most often seen the disposition of this supposed surplus merely as a problem in marketing, they have advocated buying more bales of cotton, or selling more grain abroad, or curtailing production by various means. This view of production and distribution is an outgrowth of our current business psychology, at best a so-called practical but a short-range solution. It does not reach the more fundamental considerations of proper distribution (distribution as defined in economic theory), of making it possible for more men and women to have the food necessary to good health. The problem of how to achieve proper distribution, as conceived in its correct economic way, is not peculiar to agriculture but to all industry.

Co-operative marketing has been offered as a panacea for agricultural marketing ills. There are many excellent arguments favorable to the establishment of co-operative marketing centers in certain areas and with certain products; but in the opinion of many it is unwise to advocate their general establishment. Although the conventional "middleman" has often extracted an unduly large proportion of the gross returns for his services, he has rendered a valuable service to the farmer and has often rendered this service at a lower cost than the co-operative association could render it. Marketing of many farm products in America is vitally affected by the great distances through which they must be transported. Marketing of some products involves purely domestic problems, whereas the marketing of others involves world problems. No one general type of marketing organization has thus far successfully met these extremely complex American problems. America employs both the "co-operative" and the "middleman" organizations.

6. *Credit and Income*

The economic welfare of those engaged in the agricultural industry is vitally affected by the credit structure. Time is an important element both in agricultural production and in credit. In very few agricultural pursuits is there a great daily or even monthly turnover of products that may be sold for cash. With the exception of some truck crops it is often necessary to wait from several months to several years for a realization of a cash return on the original seed planted in the ground. Those who depend on livestock (particularly for the production of meat) often must wait several years before they can realize a return on a given investment. Few individuals are able to carry on these undertakings over a period of time without the means of some credit.

The bulk of the short-time credit has been supplied by local private banking interests. Prior to some recent banking reforms and legislation by our state and national governments, these local banks constituted probably the weakest unit in the agricultural credit system. Short-time credit for the production of crops and animals has been needed by the agricultural producers for many years, but the close dependence of agriculture on nature has made risk an especially important element in the granting of this credit. Often, therefore, the banker has felt it necessary to charge high interest rates or to limit seriously the amount of the loan. Flood conditions, hail, drouth, insect damage, and numerous other natural interventions often make it impossible for agricultural producers in a given area to meet their financial obligations in any one year. Without the help of other credit institutions local banks have often been forced to close their doors. But with the help of state and national legislation effecting banking reforms, and with more judicious banking practice, the short-time credit needs of the farmer have been met in a much more satisfactory manner.

Mortgages are a very common form of long-time credit used in the agricultural industry to buy land and improvements. Payments on them often spread over the life-time of an individual producer. Mortgages contracted for during periods of high prices and inflation are sometimes the cause of much hardship during times of low prices and deflation. Variations in commodity prices often make it necessary to take a far larger proportion of the total crop to meet mortgage obligations than it did when they were incurred. Some of our national leaders, during the recent depression, even advocated repudiation of those mortgages which became so burdensome. There probably should be some means by which these obligations could be more nearly scaled to the earning capacity of the industry, and these means are being partially developed through various re-financing measures. But, in spite of these efforts, insurance companies and banks have come to claim a place among our major absentee landlords. There is ample evidence of need for further study of these long-time credit needs of the agricultural industry.

The amount of income realized by the agricultural industry has been a matter of great concern to many students of agricultural problems. There have been repeated statements by leaders to the effect that agriculture has not been receiving its share of the national income in proportion to the investment and the efforts. The prosperity of the agricultural industry as an operating proposition depends largely on the purchasing power of its own income in terms of the goods and services which it uses. Following the World War there was a virtual cessation of the foreign demand for many staple agricultural products. Urban industrial prices were maintained by various devices, such as the tariff and the international loans, while agricultural prices dropped measurably. In terms of the goods they could buy and the services they could command, agricultural producers were at a disadvantage. So there arose the claim of agriculture to a larger share in the national income. During recent years many measures have been advocated by our legislators to increase the agricultural income. This has been one of the avowed purposes of the AAA. Before the AAA, the Farm Board operated. Subsidies in several forms have been given. These and many other changes have been proposed to help make possible a distribution of the national income which will enable farmers to live as well as other groups of people who apply equal intelligence, skill, energy, and capital to their productive efforts.

THE IMPORTANCE OF THESE CURRICULUM AREAS AND RELATED PROBLEMS

The foregoing treatment of six curriculum areas (The National Government, Rural Population, Land, Production, Distribution, and Credit and Income) was included to give the reader a better understanding of their importance and a better comprehension of some related ideas. To be sure, the treatment here is brief and is only a sample of what might develop in classroom discussion, debates, etc., upon these areas and their related problems. As indicated in Chapter I, *problems of contemporary life should become objects of study in our public schools.* The problems cited in connection with the six curriculum areas are representative of many that are of national concern. Yet those noted are so fundamental that they are here proposed for study in all our public secondary schools, urban and rural.

HOW TO USE THE GENERALIZATIONS

Problems constitute one of the two forms of subject matter developed for the use of curriculum workers in this study. Six curriculum areas and their related problems concerning the agricultural industry have been suggested for study in all our secondary schools. But these problems need clarification. Generalizations, therefore, are presented in this study as the second form of subject matter which will assist in solving these important problems. (See Chapter III for the body of generalizations.) However, there are several ways open to the teacher for use of the generalizations; there are at least two major viewpoints from which they may be considered. One may be termed the *end.*

THE "END" VIEWPOINT IN UTILIZING GENERALIZATIONS

This viewpoint holds that the generalizations stated are the goals or the ends toward which study is directed.

In one application of this viewpoint the generalizations might be presented to the pupils as a body of factual material to be memorized. The generalizations might or might not be associated with problems. No doubt the teacher would assist the pupils in the process of memorizing and probably would assign certain additional subject material to help them in gaining a better understanding of the generalizations. Pupils would have no part in developing them; there would be very little reflective thinking on the part of either pupils or teachers; learning would be almost wholly on the *memoriter* level. To follow out this application, a teacher would take all or parts of Chapter III of this study and put the material directly into the hands of the pupils, to be memorized. It would be the main textbook, the subject matter to be assimilated; it would be the principal source material for lesson or unit assignments. Probably this application of the "end" viewpoint is the least desirable in schools, yet it might result in a fund of information that otherwise would not be in the possession of the pupils.

But a more desirable application of the *end* viewpoint could be made. Here specific generalizations would not be presented to the pupils to be memorized. Instead, the teacher would assign certain topics for study by the pupils and through this study would have them arrive at generalizations. The end of learning would still be generalizations, but in this instance the teacher alone would be aware of their nature before study had taken place. The teacher alone would have them in his possession at the beginning of the class work, whereas both teacher and pupils had them for use in the first application explained. Probably the generalizations arrived at by the pupils under the guidance of the teacher would not be unlike those known by him, yet they would be gained through more creative activity by pupils and would be stated in their terms. The teacher would have generalizations (such as those developed in this study) in printed or typed form for his own use, but not for the pupils' use. These printed generalizations would constitute the desired factual outcomes for his teaching. They would be his guides in assisting pupils in arriving at their generalizations.

THE "MEANS-TO-AN-END" VIEWPOINT

A second main viewpoint from which generalizations might be considered may be called the *means-to-an-end*. Two applications of this viewpoint are presented here.

In one application of this viewpoint, generalizations are considered as the *means of solving problems.* They are thus a vehicle to acquire and use for the purpose of arriving at the destination, which is a consideration and a solution of a problem. Some of the generalizations as stated in Chapter III might help a class to arrive at the solution of certain agricultural problems raised by teachers and pupils. Assume such a problem as, "What can be done to prevent the great losses in our state and nation because of soil erosion?" Suppose that this problem is accepted as real and vital by a group of pupils under the guidance of a teacher. In order to solve this problem, the pupils will need certain ideas, facts, concepts, generalizations. They may arrive at these by many different routes such as direct observing, reading, listening to discussions, and seeing pictures. The teacher (in this application of the means-to-an-end viewpoint) would have access to the generalizations noted in Chapter III and might even make these available to pupils. But the teacher would preferably help the pupils in the development of such worthy generalizations as: "Too large a proportion of land in clean cultivated crops is disastrous to soil fertility." Generalizations such as this one would not be developed simply for their own value (as noted in the end viewpoint), but for their value in solving a particular problem. Their relation to the solution of the problem would be consciously noted by both teacher and pupil. So generalizations would become a means to an end; they would become a means of solving a problem.

The generalizations (stated in Chapter III) are not the only ones which pupils should develop and understand in a consideration of the more important American agricultural problems. They are in one sense a group of the more important generalizations that will be arrived at by a study of the problems, and yet in another sense they are but a lead to many other equally important generalizations. The tax specialist who examines them may be disappointed because a number of important generalizations are apparently absent. These are available to those who may wish to make a further study of particular areas in specialized sources. The area (or problem) has been suggested and certain pertinent generalizations have been suggested for study. The problems and generalizations presented in this study do represent a selected group arrived at by applying a research technique to selected publications, but they are not the "last word," are not the only ones that can be accepted. If the reader assumes that the problems and generalizations arrived at in this study are complete, authoritative, beyond question of doubt, not subject to controversy, then he will come to disillusionment. Many of the generalizations presented are widely accepted as truths; many are objects of controversy. Apparently the social sciences have not reached that stage of perfection or finality where proposed solutions to problems are incontrovertible. Among other things this means that additional problems and generalizations should be developed by teachers and pupils as they work in the significant agricultural curriculum areas. They should not accept these problems and generalizations as final, but only as tentative, to be tested or used in further study. They should be considered as means-to-ends but not as ends. The principal concern of the teacher (as emphasized here) is to see that pupils do arrive at worthy generalizations in their consideration of our more contemporary problems, that they consider these generalizations as tentative, and that they look on them as understandings which are of value because they do help solve these problems.

The principal concern of the teacher, according another application of the "means-to-an-end" viewpoint, becomes largely a matter of seeing that pupils develop in situations that involve problems which are challenging, purposeful, and vitally significant. Bayles[4] accordingly gives this interesting criticism on the use of generalizations, or recognitions as he terms them.

> In such situations, the teacher is not so much concerned in getting the right answers every time, as in getting the student always to *follow out* the implications of his answers. It is not so important that "situations and responses which belong together are actually brought together" as that there be repeated opportunities for seeing answers in contrast and for thinking through to the selection of a defensible answer.

[4] Ernest E. Bayles, "Generalizing or Thinking?" *Educational Administration and Supervision,* September, 1932, pp. 438-39.

In this manner, the really important thing that the student carries away with him is this method of attack on problems and the particular recognitions gained can be permitted more or less to take care of themselves. If a particular recognition that has been missed in the instruction is needed at a later time, the student will have been trained in a method of *arriving* at dependable recognitions, and this will be available at the time of need. In other words, we will have made him independent of the teacher . . . we will have made his training transferable. . . .

USING PROBLEMS AND GENERALIZATIONS IN A COURSE OF STUDY UNIT

In order to make still more concrete a way in which the problems and generalizations may be used by teachers and curriculum workers, Chapter V presents a Course of Study Unit. A Course of Study Unit is analyzed into component parts, the parts are explained, and the explanations are applied to a particular curriculum area of agriculture.

CHAPTER V

A COURSE OF STUDY UNIT

THIS chapter purposes to show an application of some problems and generalizations (presented in Chapter III) in a course of study unit. There is a discussion of different types of units; there is also a suggested organization for a course of study unit, together with a brief explanation and development of it.

WHAT IS A UNIT?

The term *unit* is now used widely by teachers and others interested in public school curriculum development, but applications of the term vary greatly. Some of this variation is due to different conceptions of the source of unity. Thus Caswell and Campbell divide units into two major types—*subject matter* and *experience.* In the subject matter type

> . . . the dominant purpose of the unit is to develop understanding of a particular body of content—expressed either in terms of facts to be mastered or generalizations to be understood—or of a given phase of the group culture, or of an aspect of the environment.[1]

Whereas in the experience type

> . . . the initial point of orientation and the source of unity is sought in the experience of the learner and the subject matter is selected to help realize ends the individual wishes to reach.[2]

Harap[3] classifies units into seven principal types on the basis of his analysis of a large number of units on file in curriculum laboratories. Bruner classifies units most used in the social studies field as *theme, social-functions,* and *experience units:*[4]

[1] Hollis L. Caswell and Doak S. Campbell, *Curriculum Development,* p. 403. 1935.

[2] *Ibid.,* p. 404.

[3] Henry Harap, "A Survey of Courses of Study Published in the Last Two Years," *Journal of Educational Research,* May, 1935, pp. 648-49.

[4] From an unpublished manuscript by Herbert B. Bruner.

The theme unit

. . . assumes that it is possible to determine a few large "themes" or "big ideas" which have controlled, are controlling, and will control the actions of man. The "themes" that have been utilized . . . are (1) interdependence (men are becoming increasingly dependent upon the efforts of each other), (2) man's control over nature (man, wherever possible, controls nature for his own ends), (3) man's adaptation to nature (in cases where man cannot control he must, of course, adapt himself as wisely as possible), (4) migration of population (man has constantly moved from place to place in search of higher standards of living), and (5) democracy (in spite of temporary set-backs the progress of democracy has seemed to be inevitable).

These large "themes" represent five hierarchies of principles, the kinds of principles which have dominated the actions of men in the past and will continue to direct the destinies of men in the future. They are on a parity with the natural laws. In fact, they are the controlling drives in the progress of human affairs. They are the most succinct descriptions of the ways in which man operates. If we can assume that these statements are true, these "themes" should serve as very valuable selectors of experiences and subject matter.[5]

The social-functions unit

. . . is predicated upon the theory that curriculum makers should determine the major functions of social life and then set up experiences and subject matter which may be utilized by pupils in becoming more proficient in performing these functions.[6]

The experience unit

. . . is based upon the principle that the curriculum should be made up of experiences which are of real interest to the child. In its extreme application the philosophy is employed that every unit must grow out of the children's experiences and the problems they are interested in solving here and now. It has as its starting point the present life of the child and the aim is growth through better living here and now.[7]

At least one other distinction between units seems to be necessary. There are *course of study units* and *teaching units.* The

[5] *Ibid.*

[6] *Ibid.*

[7] *Ibid.*

course of study unit is generally more replete with suggestions, more detailed, and more comprehensive than the teaching unit; it is developed primarily for the use of teachers and for many possible teaching situations. On the other hand, the *teaching unit* is developed by the teacher for a particular situation; it may be written briefly or fully; it is usually the teacher's own directions for work in her particular room. These two kinds of units may differ slightly in organization, but their context is usually quite different.

SUGGESTED ORGANIZATION FOR A COURSE OF STUDY UNIT IN THE SOCIAL STUDIES

The type of unit selected for development is a major determinant in organization. The organization used here is adapted more to the *theme* or *social-functions* types described by Bruner, or the *subject-matter* type described by Caswell and Campbell. However, it should be noted that both subject matter and experiences serve as bases of unity (which is characteristic of most actual classroom teaching). The following is a suggested organization for a *Course of Study Unit in the Social Studies.*

1. The Title, together with a statement of the Controlling Themes
2. The Objectives or Purposes
3. Big Problems and Related Generalizations
4. The Overview, giving Probable Points of Emphasis in the Content
5. Content (in summary form)
6. Suggestive Activities (teacher and pupil)
    a. Approach
    b. Developmental
    c. Culminating
7. Suggestions for Evaluation (teacher and pupil)
8. Glossary of Terms
9. Bibliography—Supplies—Equipment

This suggested organization contains some of the most frequently used parts for course of study units. It makes possible the inclusion of both suggestive materials and activities for the guidance of the learner, and offers a practicable means of realizing at least one of the major purposes of the social studies:

> . . . *It is the purpose of the social studies to give to the pupils the truest and most realistic knowledge that is possible of the community, state, nation, and world—the social and physical setting—in which they live and are to live and make their way.* The social studies are to show as best they can the relation of mankind to its physical environment (social geography) and then how industry, agriculture, government, and social organizations, such as chambers of commerce, trade unions, granges, and political parties are actually carried on. . . .[8]

It is toward the fulfillment of this purpose that course of study units dealing with important curriculum areas of agriculture should be included in the curriculum of the public schools. One of the most profitable agricultural curriculum areas for study is that of *Land*, profitable because it offers so many rich opportunities for better understanding of the "relation of mankind to its physical environment. . . ."

There is a brief discussion of the curriculum area in Chapter IV, pages 70 to 72, but it is deemed of such great importance that it is here designated as the area for development in a course of study unit. The material which follows in this chapter is therefore an abbreviated illustration of the manner in which that unit might be developed.

A COURSE OF STUDY UNIT WITH LAND AS THE CURRICULUM AREA FOR EMPHASIS

1. *The Title and Controlling Themes to Be Considered*

A title for a unit should furnish a clear, interesting basis for unity. Also, the title should indicate consideration of a significant area in the pupil's environment (in social studies) couched in language that he will readily understand. Some such titles as "The Major Uses of Land in America" or "Uses That We Should Make of Our Land in Texas" might be used for this unit.

As Bruner has stated, there are certain themes which tend to

[8] The Department of Superintendence, National Education Association, *The Social Studies Curriculum*, Fourteenth Yearbook, p. 57. 1936.

dominate the several fields of work in the social studies. These themes are "controlling drives" in the lives of men and are proposed as controlling drives for the social studies curriculum. Such themes as "man's control over nature" and "man's adaptation to nature" represent the controlling themes of this unit. Man has exercised control over land by cultivating it, by terracing it, by draining it, by irrigating it, etc. There are many illustrations of control to be found in man's utilization of land, but there are also many illustrations of his adaptation. Man has learned that he can grow certain crops on certain type soils; he has learned that he can engage in profitable farming on certain lands; he has learned that depletion of the humus in soil lowers its water-holding capacity; so he has found it necessary to adapt his practices to nature. Both themes (control and adaptation) are operating in man's utilization of land and they should be recognized as big, controlling factors for this curriculum unit.

2. *The Objectives or Purposes*

Some of the objectives for a course of study unit will probably be determined in advance of the teaching situation by the teacher or the course of study writer. But the learner will have his objectives too, and the teacher should assist him in making these objectives, or purposes, worthy.

While obviously it is impossible for a group of adults to determine in advance purposes which pupils may set up for themselves, by drawing on past experiences or by envisaging the teaching and learning period when the materials are to be used, adults can suggest some of the desirable purposes which pupils are likely to choose.

One outstanding general objective a teacher has in any social studies unit is to assist pupils to draw from the activities and subject matter accurate, valuable conclusions or generalizations and translate these as far as possible into desirable action.[9]

Pupils should not only be able to derive these conclusions or generalizations, but they should also be able to identify some of the more important problems which they help to solve. Generalizations may be organized about certain big problems, themes, or

[9] *Ibid.*, p. 237.

curriculum areas of importance in order to give them greater unity and a closer relationship. Likewise, several problems may be organized about a significant generalization or generalizations. They are then more nearly functional to the pupil.

The general objective for this unit might be: *To assist pupils in arriving at some important generalizations and a better understanding of some important problems relating to the uses made of land in America.*

In addition to the large general objective . . . there are more specific purposes which may also be attained through a combination of teacher guidance, pupil activities, and content materials. These should provide direction (1) for building up proper habits of work, (2) for getting information upon which to base big understandings, and (3) for improving techniques and skills. While in the highest type of learning situation specific purposes emerge as desirable outcomes as the work progresses, it is probably wise to list . . . at the beginning of the unit some of the purposes that seem most important.[10]

As an example, specific purposes for this unit might be some of the following:

1. To learn the different major uses to which land is put in America.
2. To learn how public policy has influenced the different uses to which land has been put.
3. To learn how other major factors have influenced uses.
4. To learn about recent changes in land use and why these have taken place.
5. To study the possible land uses for a limited area near-by.
6. To develop attitudes favorable to the study of our modern problems.
7. To develop certain valuable habits of study, such as independent research for materials related to social-political-economic problems.
8. To learn how to discuss more ably some important problems in a group of fellow pupils.

[10] *Ibid.*, p. 238.

3. *Big Problems and Related Generalizations*

In the preceding section, Objectives or Purposes, *problems* and *generalizations* were cited as some desirable subject materials for the social studies. It should be an aim of the social studies teacher to try to state (discover, develop, formulate) worthy problems and generalizations, and further to guide pupils in their formulation of these. One way in which the course of study maker (or teacher) may arrive at these problems and generalizations is described in Chapter II. Several ways in which such problems and generalizations may be used are described in Chapter IV.

A few suggestions relative to the formulation of these problems and generalizations follow:

Both for administrative and psychological reasons it is advisable . . . to suggest for pupils important problems which they may explore with interest and value in connection with the work of the unit. . . . The problems should, of course, be so related to the large key idea (of the unit) . . . that their solution would lead to a better understanding of the generalizations basic to this key idea.[11]

Obviously, not all of the most valuable generalizations will occur to the course-of-study writers as a result of their prestudy or even as they write the actual materials. They should list, however, those that seem to them to be most essential. . . . Teachers subscribing to the philosophy that permits pupils and happenings of each day to determine in some measure the direction of the work will know that in many instances new generalizations will be added as the pupils work through the unit. In writing the materials, *the point should be very clearly made to all teachers that every effort should be exerted to assist pupils in arriving at the generalizations through their own activities and study.* The generalizations are placed in the course of study as a guide to teachers in helping pupils to arrive at the types of conclusions which would be of most value.[12]

Some of the probable big problems and related generalizations that are applicable to this unit on land utilization follow:

Central problems

To what different uses are our lands now put?

To what different uses should our land be put?

[11] *Ibid.*, pp. 238-39.

[12] *Ibid.*, pp. 237-38.

Some related problems

What are the principal factors determining the use to which land is put?

What effect will population increase or decrease have on land use?

What should be our national policy with relation to the development and settlement of our agricultural land?

What disposition should be made of the submarginal agricultural land?

How can we maintain the productive power of our agricultural land?

What effect does tenancy have on land use?

The following are some generalizations (first presented in Chapter III) which relate to the above suggested problems. In one sense these generalizations may be thought of as outcomes from a study of the unit, but they are probably of most value when developed by pupils and used by them as aids in the solution of problems (see the explanation of the means-to-an-end viewpoint set forth in Chapter IV). This list of generalizations is longer than the usual class develops.

Generalizations

The first step in putting each kind of land to its proper use is a scientific economic survey of the land.

A given region's available land uses are more likely to be complementary than competitive, and the common interest is best served when all the promising associated uses are encouraged to function to the fullest capacity of our social need.

Land classification for uses involves many political considerations and affects vast business interests.

Land economists have repeatedly said that there is too much land in agricultural use.

We have come to the time when we should write a new epic—the epic of adjustments, of regrouping, of retirement from cultivation of lands which the pioneer subdued but which now refuse to yield a reasonable scale of living.

Not all soils and places where farms are located are adapted to farming.

A study of land resources and future requirements for farming should be answered in terms of an economic organization pattern; that is, in terms of types of farming in given regions.

Much land formerly in crops should now be put in grass or forests.

The utilization of available farm land resources should be directed as between farming uses proper, grazing, and forestry, in keeping with the interests of established farmers and the needs of the nation.

Although all classes of economic productive activity require some land, agriculture makes the greatest demands for this agent of production.

Regional planning related to water supplies, flood control, residential development, and recreational facilities should be co-ordinated with the rational use of the land for farm purposes.

There should be a development of our national forests and other public reservations designed to utilize land for public welfare.

There is a great variation in the economic situation of land; that is, its situation with relation to cities, roads, and economic institutions.

Agriculture has suffered from a haphazard development of the land.

It is desirable that a better co-ordination be effected among the various national and state agencies concerned with the administration of public lands in order that there may be unified action in bringing about the most economical use of our land resources.

In our national reclamation policy it seems highly desirable to weigh the advantages of local or regional development against the disadvantages of promoting excessive agricultural expansion.

Broad considerations of economic, social, and military policy may justify reclamation as a means of increasing the land areas, but the United States is not in a position to justify this means.

The national government has been operating under land laws that were suitable to a self-sufficing agriculture but are out of place in a commercial agriculture.

Land legislation that expands the agricultural area may at times be justified, but at other times it is of great importance to provide the means of taking land out of agricultural use and helping the people who are stranded on submarginal land to move to regions and occupations which will provide the opportunity to make a decent living.

Political units should consider the advisability of acquiring privately owned land, which has been abandoned or which is not suitable for farming, and using it for forestry and recreation.

There should be a system of grazing control for the public domain outside of the national forests, comparable with and unified with that now in force in the national forests.

Public ownership will always play a large role in forestry.

Our past agrarian policies have been responsible for a continuous over-expansion of agricultural lands.

The policy of land settlement in this country has been to divide the land among as many people as possible.

In general, since the formation of our government, there have been few restrictions on the use of land for agriculture.

The over-expansion of agriculture and the bringing of marginal land into use are not so much due to the natural land hunger of the people as to the stimulation of various interested agencies.

The liberal disposal of the great body of semi-arid land in the western half of the United States has added to our land troubles.

One of the most striking features of the war development of agriculture was its spread into comparatively low-grade lands with fairly extensive methods of use.

Most of the expansion of our agricultural lands takes place in four directions: drainage, irrigation, clearing forests, and dry land farming.

We have in cultivation a vast area of submarginal land, at normal exchange ratios.

Submarginal land, from the social point of view, is land from which the average farmer cannot realize a satisfactory standard of living.

If submarginal lands were withdrawn from production by private interests under favorable legislation or placed under the control of state or national agencies to be used for the purposes for which they are best suited, at the same time conserving the soil, a great step forward would be taken.

Favorable taxation and credit policies would tend to promote the use of inferior lands now in farms or likely to be thrown into farms, for the growing of forests and pastures.

As a nation we possess extensive land resources, yet in no other nation is the depletion of soil resources taking place so rapidly.

The era of land abundance has gone.

It is seldom that one generation suffers the full penalty of soil loss; the top soil vanishes too slowly to make its conservation a constant and immediate aim in cultivation.

All people share with the farmers the responsibility for maintaining the producing power of the land.

4. *The Overview (Giving Probable Points of Emphasis in the Content)*

The *overview* should give a *point of view* to the person who contemplates teaching the unit. It should be put to much the same use as the compass is put by the mariner. By it he should be able to chart a course in a given direction. By the *overview* the teacher should be able to gain a clear idea of the probable content which should be emphasized in the unit. It should serve as a foundation blue-print on which the superstructure of the unit may be built. Such items as themes may be here related to the particular subject content area of the unit. Since this is to be an overview, it should be comparatively brief and pointed. Such an overview for this unit, related to the uses of land, follows.

We are here concerned with the physical-economic uses of land, that is, with the agricultural and non-agricultural uses made of land. The teacher should keep in mind these additional points for emphasis in the unit: (1) Land utilization is a social problem of great concern to both rural and urban people; it is not the farmers' problem alone. (2) Loss of land resources is taking place at an alarming rate in America. (3) Exploitation and individual profit were the chief characteristics of the early American land use concept; and the land use policy of the early government was mainly one of land disposal to those interested in getting it for personal benefit. (4) The modern concept of land use is one of intelligent planning for both agricultural and non-agricultural purposes, by both governmental and non-governmental agencies. (5) "The natural resources of America are the heritage of the whole nation and should be conserved and utilized for the benefit of all our people. Our national democracy is built upon the principle that the gains of our civilization are essentially mass gains and should be administered for the benefit of the many rather than of the few; our priceless resources of soil, water, minerals are for the service of the American people, for the promotion of the welfare and the well-being of all citizens."[13]

5. *Content for the Unit on "The Major Uses of Land in America"*

(This is a summary of materials for the teacher's use)

A. *Land Utilization Should Be of Concern to All*

Land and its uses should be of concern to all citizens of the

[13] National Resources Board Report, December 1, 1934, p. v (Foreword). 1935.

United States. Although we live in an industrial age and in a highly industrialized country, we are still dependent upon the land for our basic necessities of food, fuel, clothing, and shelter. Rural and urban people alike must look to the land and its products when these necessities are considered.

The waste of these natural resources has been called to our attention by recent studies on soil erosion. "The soil-erosion specialists tell us that the dust storm of May 11, 1934, swept 300 million tons of fertile topsoil off the great wheat plains; that 400 million tons of soil materials are washed annually into the Mississippi river; that generally water and wind erosion together each year remove beyond use 3 billion tons of soil.

"They find that 100 million once-fertile acres of farm land—equal to Illinois, Ohio, Maryland, and North Carolina combined—have been essentially destroyed for profitable farming; that another 125 million acres are seriously impaired; and that another 100 million acres are threatened—all belonging to the best farm lands of the United States.

"And further; that the present annual money loss to landowners and to the nation is not less than 400 million dollars each year; that the annual rate has been increasing; that the cumulative loss may be conservatively stated as already not less than 10 billion dollars; and that, if the wastage is not stopped, in another 50 years the cumulative loss will reach the staggering figures of 25 or 30 billion dollars, equivalent to a loss of $4,000 on each and every farm in the United States."[14]

The maintenance of a growing, prosperous society in America is in a large measure related to the future uses made of land. Barbarous, nomadic man was not a costly user of land, but modern, civilized man is. His plow has often been the instrument of exploitation, the mining tool which has caused the loss of great soil resources. The population of our country, and of the world, has increased tremendously in the last one hundred years and so our exploitation of land resources has gone on at a mad pace. Recent discoveries in Mexico would indicate that one of the

[14] Person, H. S., *Little Waters. Their Use and Relations to the Land,* p. 2. 1935.

principal causes for the disappearance of the then highly cultured Mayan people was their neglect of proper land utilization. One of the major contributing causes for the downfall of the Roman Empire was the loss of fertile lands. Mesopotamia, often called the cradle of civilization, no longer supports a great people, and one poignant reason is her ruined land. We in America should take heed, for: "The fact is that most of the territory occupied by the United States is not naturally suited for a permanent civilization. It is like the land of the Mayas of Yucatan or the land of Babylon—a rich country where civilization can flash into a blaze of glory and then collapse in a few generations into ruin. Our soil is not enriched by the usual methods of cultivation, but impoverished. By the normal processes of our farming, our mining, and our lumbering we create a desert."[15]

Land, therefore, is not the concern of the individual user alone, for he who uses land is in a measure the captain of society's destiny as well as of his own. Land utilization should be the concern not of farmers alone but of all classes of people.

B. *Social, Political, and Economic Implications*

Land utilization has many social, political, and economic implications, only a few of which can be enumerated here. One may think of land in terms of the *physical-economic* uses to which it may be put as an agent of production. It is from this viewpoint that we intend to develop this course of study unit. Thus we may think of land in terms of *agricultural and non-agricultural uses.* Some lands, because of their fertility, are well suited to certain kinds of agricultural uses, whereas other lands, because of their location (for example), are suited to certain kinds of non-agricultural uses.

There are numerous *political considerations* that may affect land utilization. For example, low taxation or high taxation may definitely affect the extent and kind of use; governmental zoning may determine the limits of certain uses; subsidies or bonuses may influence uses; tariffs and international trade agreements may have their influence, and so on. All are factors which may enter into the broader determination of land use.

[15] National Resources Committee Progress Report, June 15, 1935, p. 6. 1935.

But man must be considered in relation to his use of land, too. When land is occupied there are people living on it, and when it is abandoned people must move. And so arise the problems of ownership and tenancy, of occupancy and settlement, which are more nearly social-economic phases of land utilization. These and other problems relating to land tenure are very important, but not considered within the province of this course of study unit.

c. *The Early American Concept of Land Use*

The early American concept of land use may be considered from the viewpoint of (1) the individual, and (2) the government.

The *individual* considered it his privelege and opportunity to exploit the land and its resources. America was then almost wholly an agricultural user of land. Pioneers cleared the land of trees and turned the sod for the production of crops. As soon as the land showed definite signs of "wearing out," the farmer and his family turned westward with the advancing frontier and cultivated "new land." This trek across our country has continued until most of its better agricultural land has been exploited in varying degrees. Thus the virgin, fertile lands of our country have been literally "mined" for their wealth. Exploitation and individual profit were the chief characteristics of the land use concept accepted by early Americans.

The *early government's policy* with respect to land may be more properly characterized as a policy of *land disposal.* There was little thought given to desirable uses, with a few notable exceptions. Starting with the Northwest Ordinance, certain lands were set aside for the establishment of schools. Thousands of acres were given to railroads for the building of transportation lines into the new territory. The early Homestead Laws gave land to practically any person who was willing to risk settlement. There were no stipulations as to what should be done with the land. The policy was simply to divide the land among as many people as possible and get it settled as rapidly as possible. Immigrants from Europe supplied the growing nation with the many hands needed for conquering the land. So our government early encouraged private ownership. There was little thought of retaining any productive land for governmental ownership; private ownership was heightened; the individual entrepeneur was a virtual king of his domain.

D. *Changing Concepts of Land Use*

With the passing of our frontier there came a realization of great extravagance in the utilization of our land; then Theodore Roosevelt helped to usher in a national conservation movement. The government seemed to awaken to some of its responsibilities. National forests were established to conserve timber; national parks were established to conserve wild life; agricultural colleges and experiment stations were established; extension services were developed. The twentieth century witnessed the beginning of a widespread emphasis on conservation by governmental and voluntary agencies.

Soon afterward came the World War and lands were used both extensively and intensively for production of war materials. Conservation was forgotten; the cry was for production at any cost. Prairie sods were broken by plows; abandoned acres were again pressed into service; more land than ever was used by producers to help supply domestic and foreign demands. Then the War ceased; foreign producers resumed their work and American products were no longer wanted in such great quantities; immigration was throttled. The spectacle of overproduction and idle land on a wide scale was witnessed in America for the first time. Agricultural producers and others began to suffer financial reverses; a major depression hit all. Gradually the idea of collective, cooperative action gained more favor. The national and state government entered the picture, both as adviser and as regulator. There came a Farm Board, then an AAA, and various other governmental agencies concerned directly and indirectly with the use of land. Out of all these activities came more clearly than ever before the concept of "planning" for land use. No longer did the nation consider the utilization of land as a haphazard undertaking.

In accordance with Executive Order No. 6777, June 30, 1934, there was submitted to the President of the United States on

November 28, 1934, a Report of the National Resources Board. In the letter of transmittal the chairman stated: "This report carries this significance: That it is the first attempt in our national history to make an inventory of our national assets and of the problems related thereto. Moreover, for the first time it draws together the foresight of the various planning agencies of the Federal Government and suggests a method for future cooperation." Thus there was launched the concept of *national planning* under the auspices of the government. We have had various attempts at local and state planning for many years, but we are now apparently on the threshold of a more valuable, integrated type of planning on a national scale.

The concept of planning is one of intelligent adjustment and development rather than one of nonadjustment or of a fixed pattern for all time. It involves co-operative work by both governmental and non-governmental agencies, and it develops from the local areas through the state and region to the nation. Local and state planning boards have been organized all over our country. State decisions on land utilization should be derived from local studies; national recommendations should in turn be based on state and regional studies, with due regard for beneficial integration on state, regional, and national scales.

E. *Major Uses of Land in America*

What are the *major physical-economic uses* to which our land is now put? We name two: agricultural and non-agricultural. The following is an outline amplifying these two main uses of land:

1. Agricultural uses of land in America
   - *a.* Crop land
   - *b.* Pasture or grazing land
   - *c.* Forest or timber land (including some pasture)
   - *d.* Land for farm buildings, lanes, and yards
   - *e.* Idle or fallow land
2. Non-agricultural uses of land in America
   - *a.* Roads and highways
   - *b.* Railroads

c. Parks and recreation centers; public domain
d. Cities and villages; cemeteries
e. Rocky, barren wastes; swamplands, marshes, beaches
f. Schools; institutions
g. Wild life preserves
h. Watersheds, reservoirs, power development; flood control
i. Industrial sites; factories
j. Reservations (Indian, etc.)

"A little over half of the land in the Nation is in farms. Of this land in farms, 38 per cent was in crops in 1929 (including crop failure), 37 per cent was in pasture (excluding woodland pasture), and 15 per cent in woodland, the remainder being crop land lying idle, farmsteads, lanes, and waste land. All crop land is in farms, but the acreage of pastures, including range land outside of farms, exceeds that in farms . . . (the land surface of the United States is in round figures 1,903 million acres). . . . The 53 million acres of land used for non-agricultural purposes and nonforest purposes is small, but its value is great, particularly the urban land. Finally there are about 77 million acres of absolute desert, rock, certain marsh lands, and coastal beaches which are now valued at almost nothing, but have a social utility for wild life and recreational use."[16]

F. *What about Future Uses of Land in America?*

The working of local, state, and national planning boards seems to promise a basis for desirable changes in land utilization. Surely the principle of planning is worthy of our most serious consideration in establishing a way for more desirable land use. The reports of these planning boards should be studied in our schools. Furthermore, school groups should become instrumental in undertaking local or community studies of land use with the view of better understanding these reports, and the needs of the locale, the state, and the nation.

"The natural resources of America are the heritage of the whole nation and should be conserved and utilized for the benefit of all our people. Our national democracy is built upon the prin-

[16] *National Resources Board Report,* p. 11.

ciple that the gains of our civilization are essentially mass gains and should be administered for the benefit of the many rather than the few; our priceless resources of soil, water, minerals are for the service of the American people, for the promotion of the welfare and the well-being of all citizens."[17]

These are the ideals which may well serve as a guide in our study of land utilization.

6. *Suggestive Activities*

Not only suggestive content materials, but activities as well, should be noted for teacher guidance. It is through various types of pupil activity that learning takes place. . . .

> The oft-quoted statement, "we learn to do by doing," is fundamentally sound. It suggests a truth generally recognized by psychologists. But it does not follow, as is sometimes concluded, that some children are active and others are not, or that overt physical activities are of superior educational worth to intellectual activities. Rather, some children are more active than others, some schools emphasize certain types of activity and other schools other types, and the educational worth of activities is determined by the characteristics of the individual and the nature of the outcomes desired.[18]

One of the most helpful general lists of suggestive activities is to be found in a recent report of the Commission on the Social Studies.[19] Still other helpful descriptions of activities are to be found in *Curriculum Development* by Caswell and Campbell[20] and in the Yearbook of the National Society for the Study of Education, *The Activity Movement.*[21]

There are several ways of classifying activities in which pupils may engage, but one of the most helpful methods for teachers includes activities designated as follows: *Approach, Developmental, and Culminating.*

[17] *Ibid.,* p. v (foreword).

[18] Hollis L. Caswell and Doak S. Campbell, *op. cit.,* pp. 232-33.

[19] Truman L. Kelley and A. C. Krey, *Tests and Measurements in the Social Studies,* Report of the Commission on the Social Studies, IV, pp. 64-67. 1934.

[20] Hollis L. Caswell and Doak S. Campbell, *op. cit.,* Chap. IX.

[21] National Society for the Study of Education, *The Activity Movement,* Thirty-Third Yearbook, Part II. 1934.

A. *Approach Activities*

In every unit there should be a period of initiation, of discovering interests, of creating interest, of helping pupils to establish worth-while purposes. This is the approach or introduction to the main work of the unit. During the approach period the teacher will want to help pupils in selecting one or two big activities which might serve as culminating activities for the unit. Some possible mediums for approach to this unit on Land Utilization are suggested here.

1. *Pictures.* If the school has a motion picture projector, there are a number of films available, showing national parks, game preserves, farms, industrial centers, and the like. In case there is no motion picture projector available, slides might be shown on the same topics. There is an abundance of pictures which might be clipped from agricultural and industrial magazines and distributed to the class. Several encyclopedia publishers have included interesting pictures in their volumes. The U. S. Department of Agriculture has Farmers' Bulletins available, and Colleges of Agriculture and the U. S. Department of Agriculture have soil survey maps for distribution.

2. *Reading materials.* There are many current discussions in newspapers, magazines, and reports available for students and teachers. If possible, consult the *Agricultural Index* for suggestions on recent articles on Land Utilization. Assignments could be made in some of the more interesting sources, or the teacher might read to the class from these sources.

3. *Interviews, talks.* In county seats the county agricultural agent would be available. Other people who might be consulted are: teachers of vocational agriculture; members of land planning boards; college instructors in soils or agronomy; agricultural representatives of railroads, banks, implement companies, fertilizer companies, etc.; farmers; representatives of the U. S. D. A.; Federal Farm Loan officials; landowners, tenants.

4. *Excursions, field trips.* It might be possible to visit near-by farms, parks, forests, industrial sites, drainage areas, reservoirs, and the like, to obtain direct observation of different uses to which land is put.

B. *Culminating Activity*

The activity or activities selected for the culmination of the unit and the approach activity or activities have a decided influence on the developmental activities; hence the culminating activity is discussed here, somewhat out of the logical sequence. The culminating activity should be one that includes all the pupils in the class, that calls for a variety of useful preparatory or developmental activities, that seems to afford an opportunity for summarizing and bringing to a climax the previous activities dealing with the unit, and that makes possible a final, full realization of the purposes established for the unit. There are at least two major activities, either of which might well serve as a culminating activity for this unit on land utilization. They are as follows:

1. *A land planning board meeting.* Part of the class might organize as a land planning board, and part as people interested in appearing before the board. The culminating activity might be a meeting of the board at which time various people would appear before it. These people (members of the class) could present charts, maps, graphs, reports, pictures, and other data for consideration. They could represent farmers, business men, legislators, women's clubs, agricultural experts, people desiring parks, people interested in power development, and so on. They would appear before the board to discuss important problems relating to land utilization. Such a meeting could be staged in an assembly or in the classroom.

2. *An exhibit.* This exhibit might be divided into a number of sections, as (*a*) different uses to which our lands are now put, (*b*) different uses to which our land should be put, (*c*) some practices that are wasting our lands, (*d*) some practices that are conserving our lands, (*e*) illustrations of submarginal land, together with possible changes, (*f*) our great food- and cloth-producing regions shown on animated maps, (*g*) farm land in our state fifty years ago and now, (*h*) a model agricultural-industrial use of land showing complementary uses.

Either of these suggested culminating activities could engage an entire class in many and varied worth-while activities.

c. *Development Activities*

Developmental activities, as the name would indicate, are those engaged in after the approach has been made and prior to the major or culminating activity. They should be of value in achieving the purposes of the unit and be somewhat in the nature of a preparation for the culminating activity. They will vary greatly, and it would be difficult to suggest specific activities without knowing a number of factors pertinent to a particular teaching situation. But there are a number of types of activities which might be used, such as: reading textbooks, magazines, bulletins, newspapers, monographs, and letters for information; writing letters; making surveys; taking field trips; interviewing people; studying and making maps, charts, and pictures; giving oral reports; listening to talks; debating subjects; making outlines and digests of articles; constructing models; and dramatizing events. Pupils will be interested in suggesting different activities suitable to achievement of the purposes agreed on and to the needs of the particular teaching-learning situation. The teacher acts mainly in the capacity of guide and counselor, helping pupils to select desirable and valuable activities.

Organizing Pupil Activities About Problems

In place of organizing in terms of approach, developmental, and culminating activities, the teacher may desire to organize activities around stated problems. For example, the teacher might wish to organize activities around such a question as this:

Problem: What are the present major uses of land in America?

Suggested Pupil Activities:

(1) Read the references noted.
(2) Make a topical outline, setting forth the major uses.
(3) Develop a "pie graph" on plain 8½ × 11 paper showing the major uses of land in proportion to the acreage involved.
(4) Write down at least five conclusions or generalizations which you have reached with respect to present major uses of land in America.
(5) Discuss the question in round-table fashion with the class, using your outline, graph, and conclusions as bases.

Organizing Pupil Activities About Generalizations

Another way of organizing pupil activities is around pertinent generalizations. This basis was suggested for use in the sciences by Dr. S. Ralph Powers and others.[22] In accordance with this procedure, a series of generalizations, or sometimes one generalization, might be given to the pupils, who would engage in appropriate activities in order to acquire a more adequate understanding and appreciation of them. For example, the following generalization might be selected for development in connection with a unit on land utilization:

> Regional planning related to water supplies, flood control, residential development, and recreational facilities should be co-ordinated with the rational use of the land for farm purposes.

Suggested pupil activities to gain a better understanding of this generalization might be:

1. Read the suggested references.
2. Prepare a glossary of the following terms: region, planning board, land, etc.
3. Consult a regional map prepared by the State Planning Board and note the different major uses made of the land.
4. Read the Report of the TVA plans and compare these with the development in Blank river valley region.
5. Pick out any locality subject to recent floods and try to discover reasons for the same. Try to discover possible relationships of flood control to land use.
6. Statistics show that the ground water table in the middle western states (Iowa, Illinois, Nebraska, etc.) has dropped many feet in the past fifty years. Note the changes in farming practices and try to discover a possible relationship.
7. Hold a panel discussion on the topic "Regional Planning for Land Use."

Three type illustrations have been presented, showing three ways in which the problems and generalizations (presented on

[22] National Society for the Study of Education, *A Program for Teaching Science*, Thirty-First Yearbook, Part I. 1932.

pages 89-91) may be treated in relation to pupil activities. All three types are used in our better school systems. An evaluation of the three is not presented here but is left to the individual desiring to make use of the materials.

Assisting pupils to choose the most promising activities from the large number which may be suggested is one of the chief tasks of the social studies teacher. To be most effective, activities which are likely to be closely tied up with the purposes of the pupils themselves should be proposed. If this is not done, there is a very real danger that the work may become as formal as any teacher-assigned block of straight subject matter. There is no reason, however, why the wise teacher should not utilize those interests of pupils which are connected with the large social and economic problems that most vitally affect the lives of men, women, and children.[23]

7. *Suggestions for Evaluation*

There has always been some evaluation of school work by both teachers and pupils, but this has been too often concerned with information or facts alone. There is need for other evaluations. Some excellent suggestions for procedures are given in the Fourteenth Yearbook of the Department of Superintendence.[24] There are suggested some thirteen different items which may be appraised in order to evaluate the growth of pupils through the social studies.

1. Acquisition of important information.
2. Familiarity with technical vocabulary.
3. Familiarity with dependable sources of information on current social issues.
4. Immunity to malicious propaganda.
5. Facility in interpreting social science data.
6. Facility in applying significant facts and principles to social problems of daily life.
7. Skill in investigating social science problems.
8. Interest in reading about social problems and in discussing them.

[23] Fourteenth Yearbook, Department of Superintendence, *The Social Studies Curriculum,* p. 239.

[24] *Ibid.,* Chap. III.

9. Sensitivity to current social problems.
10. Interest in human welfare.
11. The habit of working co-operatively with others.
12. The habit of collecting and considering appropriate evidence before making important social decisions.
13. Attitudes favorable to social improvement.

Note that information is only one of the thirteen items for evaluation.

As another procedure, evaluation might be in terms of the general aims of education or generalized controls of conduct. Such a procedure is advocated by Caswell and Campbell.

> When the aims of education are stated only in terms of specific knowledge or habits, the types of tests usually found in courses of study may be considered a reasonably satisfactory means of evaluating instruction. However, when generalized controls of conduct are expected as outcomes, other means of evaluation must be provided.[25]

The latter means more evaluation of abilities, habits, appreciations, and attitudes, which may be measured partly by an observation type test.

Closely related to this type of evaluation is that which is in terms of the objectives set up at the beginning of a unit and during the development of it. Thus the evaluation of this unit might be in terms of the objectives set up on page 87 and such additional objectives as might be chosen with the development of the unit.

There are at least two other observations which should be taken into account in the evaluation of units. There is need for (1) both teacher and pupil evaluation, and for (2) continuous evaluation. Too often the teacher alone makes an evaluation, when there is also need for self-evaluation by the pupil. Pupils will need suggestions and some should be given in the course of study unit to assist the teacher in his guidance of pupils. Again, too often evaluation has been thought of as a test which comes at the close of the unit. If there is to be a desirable pupil growth and if evaluation is to be a help rather than a hindrance, there must be provision for evaluation throughout the development of the unit rather than just at its termination.

[25] Hollis L. Caswell and Doak S. Campbell, *op. cit.*, pp. 370-71.

8. *Glossary of Terms*

In every social studies course of study unit there will be certain terms which will need definition or interpretation (such terms will occur particularly in the Content and in the Generalizations). The glossary is an important part of the unit for giving clearer meaning to such terms. For example, in this unit on Land, the following terms should be among those included in a glossary:

Fertility: the capacity of a soil to produce crops under favorable conditions.

Land: includes all natural agents of crop production such as the air, the sunshine, the rainfall, the windblow, and the soil.

Rent: the amount paid for the annual use of land.

9. *Bibliography, Supplies, Equipment*

This part of the unit might well be stated for the two main participants, pupil and teacher. There will be certain references that are particularly good for reading by the pupils, whereas others will be useful to teachers only. The bibliography should be as complete as possible, noting a variety of worthy source materials. It should also be exact in page references when only parts of the publication apply to the unit.

Any supplies or equipment which might prove helpful in the development of activities could be referred to in this part of the unit. Sources of supply and probable costs are items that would be useful to the teacher.

The following is a short bibliography of publications that should prove interesting on land use, and that should not be difficult to obtain from the respective publishers or from libraries:

1. *Pupil*

Gray, Lewis C. *Land Planning.* Public Policy Pamphlet, No. 19. Chicago: The University of Chicago Press, 1936. 37 p.

Land Utilization and State Planning. Unit Study Booklet, No. 19. Modern Problem Series. Columbus, Ohio: American Education Press, Inc., 1933. 24 p.

National Resources Board Report, December 1, 1934, pp. 2-4; 8-25; 89-251. Washington, D. C.: Government Printing Office, 1934.

National Resources Committee Progress Report with Statements of

Coordinating Committees, June 15, 1936, pp. 6-10; 25-26; 31-33. Washington, D. C.: United States Government Printing Office, 1936.

2. *Teacher*

(In addition to the foregoing publications teachers may well read the following, some of which would also be very good for pupil use.)

James, Harlean. *Land Planning in the United States for City, State and Nation.* New York: The Macmillan Company, 1926. 416 p.

Proceedings of the National Conference on Land Utilization. Chicago, Illinois, Nov. 19-21, 1931. Washington, D. C.: Government Printing Office, 1932. 251 p.

Proceedings of Land Use Symposium. Summer Meeting American Association for the Advancement of Science at Syracuse University—Syracuse, New York, Tuesday, June 21, 1932.

Further references may be secured from the following source: *Land Utilization and Land Policies in the U. S.* preliminary list of references, compiled by L. O. Bercaw, under the direction of M. G. Lacy. U. S. Bureau of Agricultural Economics, Washington, D. C. 1936. 34 p.

CHAPTER VI

SUMMARY

The American agricultural industry is confronted with many problems of such great importance that all citizens should have a more adequate understanding of them; and all citizens will have the opportunity to acquire a more adequate understanding of them when they are considered in the public schools.

The purpose of this study has been to present, for use in secondary schools (high schools), certain curriculum materials that may prove helpful to teachers and students of the social studies. The materials presented include the more important and persistent agricultural problems, organized into three classifications: the social, the political, and the economic. Together with these problems are found some important related generalizations. Stated problems and generalizations, then, are the basic curriculum materials submitted as the principal contribution of this study for a more adequate understanding of the important problems confronting our American agricultural industry.

The materials will be of value to social studies teachers in secondary schools who may desire to base some of their work on the more persistent contemporary problems of society. It is here contended that these problems should be studied in schools. Furthermore, both urban and rural schools should give consideration to the general or non-vocational agricultural problems, such as are here presented. This study is not so much designed to help agricultural teachers as it is to help *social studies* teachers who may not have had training in the field of agriculture. This is likewise a contribution not in the field of vocational education, but in the field of general education. Agricultural problems (particularly those which are of the non-vocational or general type) should be considered a part of the social studies program in our secondary schools.

In Chapter II the technique is described whereby the agricultural problems and generalizations were derived. Guiding definitions of and criteria for determining both problems and generalizations are set forth. Current magazines, books, and national reports are indicated as sources for the stated materials. The manner of selecting each publication for analysis is stated. Then the criteria for analysis of source materials and the steps followed in the analysis are related.

Chapter III contains the problems and generalizations developed as a result of the analysis. For purposes of organization the problems are placed in three categories: the political, the social, and the economic. Under each problem there are certain pertinent, related generalizations that help one to understand it and its solution better. No one viewpoint—conservative, progressive, or radical—is maintained throughout the generalizations. On the contrary an attempt has been made to include different views as they were discovered in the source materials that were analyzed.

Out of the great number of problems submitted for possible consideration there are six groups which are selected as of probable universal interest: the National Government, Rural Population, Land, Production, Distribution, Credit and Income. A brief account of possible subject materials for each of these (curriculum) areas is given for the benefit of curriculum workers.

Several ways in which the materials (problems and generalizations) may be used by curriculum workers are presented in Chapter IV. Two principal viewpoints on the use of the generalizations, the "end" and the "means-to-an-end," are developed, together with a suggestion of preference for the latter.

In order to make still more concrete the ways in which the curriculum materials (problems and generalizations) may be used, a course of study unit is developed in the fifth chapter. Units of various types are now a popular means of organizing subject matter and pupil experiences for classroom work. So various types of units are defined, and a suggestion is made for a particular unit organization to be used in the social studies. Such a unit is developed around the problem of the uses to which our land may be put. Suggestions for the incorporation of both subject matter and activities, or experiences, make the unit adaptable to practice in schools emphasizing either "subject matter" or "experience" type units.

BIBLIOGRAPHY

American Farm Problem. Selected list of books published in the United States on the economic status of the farmer and measures for his relief since 1920. Washington: U. S. Bureau of Agricultural Economics.

Bailey, Liberty H. *The Harvest of the Year to the Tiller of the Soil.* New York: The Macmillan Company, 1927. 209 p.

Bayles, Ernest E. "Generalizing or Thinking?" *Educational Administration and Supervision,* 18:438-49; September, 1932.

Beard, Charles A. and Beard, Mary. *The Rise of American Civilization.* New York: The Macmillan Company, 1930. 2 V. 824, 282 p.

Billig, Florence G. *A Technique for Developing Content for a Professional Course in Science for Teachers in Elementary Schools.* Contribution to Education, No. 397. New York: Bureau of Publications, Teachers College, Columbia University, 1930. 101 p.

Billings, Neal. *Determination of Generalizations Basic to the Social Studies Curriculum.* Baltimore: Warwick & York, Inc., 1929. 289 p.

Black, John Donald. *Agricultural Reform in the United States.* New York: McGraw-Hill Book Company, 1929. 380 p.

Bobbitt, Franklin and Others. *Curriculum Investigations.* Supplementary Educational Monographs, No. 31. Chicago: University of Chicago, Department of Education, 1926. 204 p.

Bobbitt, Franklin. *How to Make a Curriculum.* New York: Houghton Mifflin Company, 1924. 292 p.

Bohlman, Edna McCaull. *A Course of Study Unit on Risk and Insurance.* A mimeographed unit produced in the Curriculum Laboratory, Teachers College, Columbia University, New York, 1935. 134 p.

Bonser, F. G. *The Elementary School Curriculum.* New York: The Macmillan Company, 1920. 466 p.

Briggs, Thomas H. *Curriculum Problems.* New York: The Macmillan Company, 1926. 138 p.

Briggs, Thomas H. *Secondary Education.* New York: The Macmillan Company, 1933. 577 p.

Bruner, Herbert B. *Some Suggestions for the Study of Modern Problems.* New York: Bureau of Publications, Teachers College, Columbia University, 1934. 66 p.

Bruner, Herbert B. *The Place of Units in Course of Study Construction.* Mimeograph material, Curriculum Laboratory, Teachers College, Columbia University, New York, 1930.

Brunner, Edmund DeS. and Kolb, John Harrison. *Rural Social Trends.* (Recent Social Trends in the United States. Monographs.) New York: McGraw-Hill Book Company, 1933. 386 p.

Brusse, Bun Bates and Ayer, Fred C. *An Activity Program in Action.* Dallas: Banks Upshaw and Company, 1935. 197 p.

Caswell, Hollis L. and Campbell, Doak S. *Curriculum Development.* New York: American Book Company, 1935. 600 p.

Charters, Werrett Wallace. *Curriculum Construction.* New York: The Macmillan Company, 1923. 352 p.

Chase, Stuart. *Government in Business.* New York: The Macmillan Company, 1935. 296 p.

Chicago, Illinois. *Report of the Survey of the Schools of Chicago, Illinois.* Institute of Educational Research, Division of Field Studies, Teachers College, Columbia University, G. D. Strayer, Director, Volume III. New York: Bureau of Publications, Teachers College, Columbia University, 1932. 227 p.

Clark, Marion G. "The Engineering of Problem Solving." *Education Survey,* 2:201-07; December, 1932.

Clay, Cassius M. *The Mainstay of American Individualism.* New York: The Macmillan Company, 1934. 269 p.

Cox, Philip W. L. *Curriculum Adjustment in the Secondary School.* Philadelphia: J. B. Lippincott Company, 1925. 306 p.

Craig, G. S. *Certain Techniques Used in Developing a Course of Study in Science for the Horace Mann Elementary School.* Contribution to Education, No. 276. New York: Bureau of Publications, Teachers College, Columbia University, 1927. 73 p.

Davenport, Eugene. *The Farm.* New York: The Macmillan Company, 1927. 462 p.

De Garmo, Charles. *Principles of Secondary Education,* Vol. II. New York: The Macmillan Company, 1908. 200 p.

Dewey, John. *Democracy and Education.* New York: The Macmillan Company, 1916. 434 p.

Dewey, John. *How We Think.* New York: D. C. Heath Company, 1910. 224 p.

Dewey, John. *School and Society.* Chicago: University of Chicago Press, revised edition, 1915. 164 p.

Draper, Edgar Marion. *Principles and Techniques of Curriculum Making.* New York: D. Appleton-Century Company, 1936. 875 p.

Engberg, R. C. *Industrial Prosperity and the Farmer.* New York: Brookings, 1927. 286 p.

Everett, Samuel and Others. *A Challenge to Secondary Education.* New York: D. Appleton-Century Company, 1935. 353 p.

F. E. Compton and Company. *Teacher's Handbook.* Compton's Pictured Teaching-Unit Materials. Chicago: 1934, 1935.

F. E. Compton and Company. *Unit on Trade.* Compton's Pictured Teaching-Unit Materials. Chicago: 1935.

Gray, Lewis C. *Introduction to Agricultural Economics.* New York: The Macmillan Company, 1924. 566 p.

Gray, Lewis C. *Land Planning.* Public Policy Pamphlet, No. 19. Chicago: The University of Chicago Press, 1936. 37 p.

Gustin, Margaret and Hays, Margaret L. *Activities in the Public School.* Chapel Hill: The University of North Carolina Press, 1934. 290 p.

Harap, Henry. "A Survey of Courses of Study Published in the Last Two Years." *Journal of Educational Research,* pp. 648-649; May, 1935.

Harlean, James. *Land Planning in the United States for City, State, and Nation.* New York: The Macmillan Company, 1926. 416 p.

Hatch, Roy W. "America's Social and Economic Problems and the Junior High School Curriculum." *Educational Outlook,* 172-176; March, 1935.

Heckert, J. W. *The Organization of Instruction Materials with Special Relation to the Elementary School Curriculum.* Contribution to Education, No. 97. New York: Bureau of Publications, Teachers College, Columbia University, 1917. 106 p.

Hockett, John A. *A Determination of the Major Social Problems of American Life.* Contribution to Education, No. 281. New York: Bureau of Publications, Teachers College, Columbia University, 1927. 101 p.

Hopkins, L. Thomas. *Curriculum Principles and Practices.* New York: Benjamin F. Sanborn Company, 1929. 617 p.

Hughes, H. D. and Others. *A Statement of Objectives for Agriculture.* Formulated by the Committee on Agricultural Philosophy and Objectives of Iowa State College, and approved unanimously by the Agricultural Staff of the College. Ames, Iowa: Iowa State College of Agriculture and Mechanical Arts, 1933.

Judd, Charles H. *Problems of Education in the United States.* A monograph published under the direction of the President's Research Committee on Social Trends. New York: McGraw-Hill Book Company, 1933. 214 p.

Kelly, Truman L. and Krey, A. C. *Tests and Measurements in the Social Sciences.* Report of the Commission on the Social Studies, American Historical Association, Part IV. New York: Charles Scribner's Sons, 1934. 635 p.

Kilpatrick, W. H. *Education and the Social Crisis.* New York: Liveright, Inc., 1932. 90 p.

Kirkpatrick, Ellis Lore. *The Farmer's Standard of Living.* New York: The Century Company, 1929. 299 p.

Land Utilization and State Planning. Unit Study Booklet No. 19. Modern Problem Series. Columbus, Ohio: American Education Press, Inc. 1933. 24 p.

Land Utilization and Land Policies in the U. S. Preliminary list of references, compiled by L. O. Bercaw, under the direction of M. G. Lacy. Washington, D. C.: U. S. Bureau of Agricultural Economics, 1936. 24 p.

Lee, Baldwin. *Issues in the Social Studies.* Lincoln School Social Science Monographs, No. 3. New York: Bureau of Publications, Teachers College, Columbia University, 1928. 191 p.

Lerrigo, Marian Olive. *Health Problem Sources.* Contribution to Education, No. 224. New York: Bureau of Publications, Teachers College, Columbia University, 1928. 191 p.

Lord, R. *Men of the Earth.* New York: Longmans, Green & Company, 1931. 298 p.

Meltzer, Hyman. *Children's Social Concepts.* Contribution to Education, No. 192. New York: Bureau of Publications, Teachers College, Columbia University, 1925. 91 p.

Meriam, J. L. *Child Life and the Curriculum.* New York: World Book Company, 1920. 538 p.

Miller, Irving Elgar. *The Psychology of Thinking.* New York: The Macmillan Company, 1909. 303 p.

National Resources Board Report, December 1, 1934. Washington, D. C.: United States Government Printing Office, 1935. 455 p.

National Resources Committee Progress Report with Statements of Coordinating Committees, June 15, 1936. Washington, D. C.: United States Government Printing Office, 1936. 61 p.

National Society for the Study of Education. *A Program for Teaching Science.* Thirty-first Yearbook, Part I. Bloomington, Ill.: Public School Publishing Company, 1932.

National Society for the Study of Education. *The Activity Movement.* Thirty-third Yearbook, Part II. Bloomington, Ill.: Public School Publishing Company, 1934. 315 p.

National Society for the Study of Education. *The Foundations and Technique of Curriculum-Making.* The Twenty-sixth Yearbook, Part II. Bloomington, Ill.: Public School Publishing Company, 1926. 210 p.

National Society for the Study of Education. *The Social Studies in the Elementary and Secondary School.* The Twenty-second Yearbook, Part II. Bloomington, Ill.: Public School Publishing Company, 1923. 344 p.

Norton, John K. and Norton, Margaret Alltucker. *Foundations of Curriculum Building.* New York: Ginn and Company, 1936. 599 p.

Person, H. S. *Little Waters: Their Use and Relation to the Land.* Washington, D. C.: United States Government Printing Office, 1936. 82 p.

Proceedings of Land Use Symposium. Summer Meeting American Association for the Advancement of Science at Syracuse University, Syracuse, New York. Tuesday, June 21, 1932.

Proceedings of the National Conference on Land Utilization. Chicago, Ill., November 19-21. Washington, D. C.: United States Government Printing Office, 1932. 251 p.

Report of the Committee on Recent Economic Changes of the President's Conference on Unemployment. Vol. 2, Chap. 8, pp. 547-602. New York: McGraw-Hill Book Company, 1929.

Report of the Conference on Economic Policy for American Agriculture. Chicago: University of Chicago, 1932.

Report of the President's Agricultural Commission, 1924. Washington: Senate Document No. 190, 68th Congress, Second Session.

Report on the Agricultural Situation by the Special Committee of Land Grant Colleges and Universities. Chicago, 1927. Washington, 1932.

Rugg, Harold and Hockett, John. *Objective Studies in Map Location.* Social Science Monographs, No. 1, Lincoln School. New York: Bureau of Publications, Teachers College, Columbia University, 1925. 132 p.

Sanderson, Dwight and Others. *Farm Income and Farm Life.* Chicago: University of Chicago Press, 1927. 324 p.

Seligman, Edwin R. A. *Economics of Farm Relief.* New York: Columbia University Press, 1929. 303 p.

Sims, Newell LeRoy. *Elements of Rural Sociology.* New York: Thos. Y. Crowell Company, 1928. 698 p.

Spillman, W. J. *Balancing the Farm Output.* New York: Orange Judd Publishing Company, 1927. 126 p.

Taylor, Carl C. *Rural Sociology.* New York: Harper & Brothers, 1926. 511 p.

Taylor, Henry C. *Outlines of Agricultural Economics.* New York: The Macmillan Company, 1925. 610 p.

The Agricultural Problem in the United States. New York: The National Industrial Conference Board, 1926. 157 p.

The Condition of Agriculture in the United States and Measures for Its Improvement: A Report by the Business Men's Commission on Agriculture. New York: National Industrial Conference Board, 1927. 273 p.

The Department of Superintendence of the National Education Association of the United States. *The Social Studies Curriculum.* Fourteenth Yearbook. Washington, D. C., 1936. 478 p.

Thorndike, E. L. *Educational Psychology* (Briefer Course). New York: Teachers College, Columbia University, 1922. 442 p.

Wallace, Henry C. *Our Debt and Duty to the Farmer.* New York: The Century Company, 1925. 232 p.

Warren, G. F. and Pearson, F. A. *The Agricultural Situation: Economic Effect of Fluctuating Prices.* New York: John Wiley and Sons, 1924. 306 p.

Williams, James Mickel. *Our Rural Heritage: The Social Psychology of Rural Development.* New York: Alfred Knopf Company, 1925. 246 p.

ADDENDA TO THE BIBLIOGRAPHY

The following are some good magazine articles selected from the group analyzed for the Problems and Generalizations, stated in Chapter III.

Agricultural Engineering

11:268-70, Aug. 1930. Engineers, farmers, and tomorrow. W. McMillan.

12:301-5, Aug. 1931. Electricity and the agriculture of the next ten years. E. A. White.

13:3-5, Jan. 1932. Mechanization, management, and the competitive position of agriculture. M. L. Wilson.

Agricultural Review

16:9, Nov. 1923. Present conditions of and future outlook of agriculture. W. K. James.

18:2, May 1925. Some long-view farm policies. H. Hoover.

21:8, Dec. 1928. Surveys the farm situation; national grange covers much ground in its annual statement.

American Economic Review

16: Sup. 134-55, March 1926. National agricultural policy. J. D. Black.

16: Sup. 156-66, March 1926. Our rural population debacle. C. C. Taylor.

18:405-27, Sept. 1928. McNary-Haugan movement. J. D. Black.

American Journal of Sociology

32:771-86, March 1927. Rural mind: a study in occupational attitude. W. C. Smith.

36:771-84, March 1931. Major trends in rural life in the United States. T. C. McCormick.

Annals of the American Academy of Political Science

117:1-304, Jan. 1924. Agricultural situation in the United States; ed. by C. L. King.

142:97-146, March 1929. Changes in production and consumption of our farm products and the trend in population. O. E. Baker.

142:97-146, March 1929. Farm relief and a permanent agriculture. R. G. Tugwell.

Country Gentleman

89:6, Oct. 1924. Competition or communism. G. E. Roberts.

89:5, Oct. 11, 1924. Crusade against monopoly. R. M. LaFollette.

89:1-2, Aug. 11, 1924. Henry Ford talks about farming. S. Crowther.

91:3-4, Jan. 1926. One great reason for farming. T. Cooper.

91:25, June 1926. Our future agricultural policy. L. J. Tabor.

92:3-5, Oct. 1927. New agriculture. S. Crowther.

92:10-11, Dec. 1927. New agriculture. S. Crowther.

92:31, March 1927. Subsistence farming. A. E. Taylor.

Current History

24:9-13, April 1926. Farmer's situation a national danger. G. W. Norriss.

26:169-74, May 1927. Distress of the American farmer. S. G. Rubinow.

29:265-81, Nov. 1928. Farm problem made clear. K. L. Butterfield, G. N. Peek, B. F. Yoakum.

33:872-6, March 1931. Farm machinery and the industrial revolution. H. V. Faulkner.

34:853-7, Sept. 1931. Agriculture's industrial revolution. W. Gard.

35:781-5, March 1932. What hope for the farmer? L. Harriss.

Journal of Farm Economics

7:102-14, Jan. 1925. Immigrants and American agriculture. A. E. Cance.

7:289-98, 299-304, July 1925. Land policy as a part of an agricultural program. G. S. Wehrwein; Discussion. B. Henderson.

7:305-24, July 1925. Place of taxation in a constructive agricultural policy. E. Englund.

7:176-95, April 1925. Relation of co-operative marketing to the adjustment of agricultural production; with discussion. O. B. Jesness.

7:153-75, April 1925. Role of public agencies in the internal readjustments of the farm. J. D. Black.

7:1-21, Jan. 1925. Some economic factors of an American agricultural policy. E. G. Nourse.

7:115-23, Jan. 1925. Tariff in relation to agriculture. J. Viner.

8:145-65, April 1926. Adjustment of agricultural production to demand. B. B. Smith.

9:401-20, Oct. 1927. American agriculture needs a new land policy. N. A. Olsen.

9:11-20, Jan. 1927. Farm problems. F. O. Lowden.

10:1-15, Jan. 1928. Which does agriculture need—readjustment or legislation? The case of legislation; with discussion. G. F. Warren.

11:357-67, July 1929. New farm economics. H. C. Taylor.

11:335-8, April 1929. Our congested frontier. T. N. Carver.

11:227-47, April 1929. Prospective displacement of the independent family farm by large farms or estate management.

11:386-401, July 1929. Significance of soil type in farm economy. I. G. Davis.

11:24-45, Jan. 1929. Who gets the benefit of improvement in agriculture? with discussion. H. E. Erdman.

13:203-30, April 1931. Outlook for land utilization in the U. S. O. E. Baker.

13:37-48, Jan. 1931. What agricultural products had we best export? O. C. Stine.

Journal of Political Economy

33:196-216, April 1925. Effects of price fluctuations on agriculture. A. H. Hansen.

34:624-41, Oct. 1926. Economic status of agriculture. H. W. Peck.

40:73-100, Feb. 1932. Farm relief, agricultural prices, and tariff. L. B. Zapoleon.

Nation

122:712, June 1926. Exploited farmer.

122:367-8, April 1926. Plight of the farmer. S. W. Brookhart.

124:192-3, Sept. 1, 1926. Tariff and the farmer. G. W. Norriss.

124:664-6, June 15, 1927. What will become of the farmer? R. G. Tugwell.

125:150-1, Aug. 17, 1927. Agriculture in the nation's economy. C. A. Beard.

127:422-3, Oct. 24, 1928. What the farmer is up against. R. Stewart.

129:745-6, Dec. 18, 1929. Prosperity, believe it or not; farmer's share. S. Chase.

134:460-2, April 20, 1932. Can the American farm be saved? E. G. Nourse.

Nation's Business

13:13-16, April 1925. Farmer is a business man; interview with W. M. Jardine. N. A. Crawford.

14:52, Nov. 1926. As business views farm problems.

15:17-19, Jan. 1297. Is there a national farm problem? J. H. Barnes.

18:28-31, Nov. 1930. Efficiency—the soundest form of farm relief. W. M. Jardine.

New Republic

42:279-80, May 1925. Economic law and the farm crisis.

45:203-5, Jan. 13, 1926. Latest edition of the farmer's revolt.

45:124-5, Dec. 23, 1925. Need for agricultural statesmanship.

47:49-50, June 2, 1926. Another farmer's defeat.

50:32-3, March 2, 1927. What next for the farmer?

53:64-7, Dec. 7, 1927. Soil waste and the flood menace. R. Stewart.

55:62-4, 90-2, June 6-13, 1928. Agricultural depression. A. P. Chew.

55:161-3, June 4, 1928. Plank on agriculture.

63:274-5, June 22, 1930. Farm board's troubles.

63:14-16, May 21, 1930. Unemployment in agriculture. A. P. Chew.

64:146-51, Sept. 24, 1930. Cross section of the rural mind. R. Lord.

67:168-70, June 1931. After two years on farm relief. A. C. Hoffman.

68:204-6, Oct. 7, 1931. Planning for agriculture; conference on economic policy for American agriculture, Chicago University. G. Soule.

70:8-10, Feb. 17, 1932, State and the land. J. C. Ransom.

Outlook

143:273, June 23, 1926. Farmer's is the nation's business.

149:255-7, June 13, 1928. Why the West will not revolt. G. M. Harger.

Political Science Quarterly

39:549-91, Dec. 1924. Problem of agriculture. R. G. Tugwell.

43:481-97, Dec. 1928. Reflection on farm relief. R. G. Tugwell.

Review of Reviews

76:45-54, July 1927. Farm problem stated. F. O. Lowden.

76:485-7, Nov. 1927. Farm relief and flood control. R. T. Ely.

76:69-71, July 1927. Farmer's tax burden. R. Stewart.

77:279-84, March 1928. Agriculture, a national problem. W. W. Head.

Rural America

5:22-6, Oct. 1927. Future of agriculture and rural life. E. C. Linderman.

6:4-5, May 1928. City must try to understand! K. L. Butterfield.

7:3-5, Dec. 1929. Government's farm program.

7:3-7, Nov. 1929. Rural organization. F. O. Lowden.

Saturday Evening Post

196:3-4, April 19, 1924. Fifty year crisis in agriculture. G. Garrett.

199:3-5, Oct. 16, 1926. Agricultural problem. W. M. Jardine.

199:12-13, April 30, 1927. Fourth age of agriculture. G. Garrett.

199:6-7, April 16, 1927. Industry, Agriculture & Co. G. Garrett.

199:3-5, April 2, 1927. McNary-Haugenism. G. Garrett.

199:7, Jan. 29, 1927. Unsolvable farm problem. E. T. Peterson.

199:6-7, June 3, 1926. Your food and our farmer. R. W. Child.

201:6-7, Oct. 6, 1928. Corn belt. G. Garrett.

201:3-5, May 11, 1929. New picture of agriculture. G. Garrett.

201:3-5, Sept. 29, 1928. Revolution in agriculture. G. Garrett.

201:10-11, Feb. 16, 1929. Some aspects of farm relief. R. Smoot.

201:37, May 4, 1929. Town comes to the farmer. W. M. Jardine.

203:57, Sept. 13, 1930. Challenge to America. C. M. Reed.

204:10-11, Dec. 5, 1931. Public attitude toward agriculture. J. F. Bell.

World's Work

50:602-13, Oct. 1925. What we can do for the farmer; a federal reserve board for the farming industry is suggested. F. O. Lowden.

51:657-61, April 1926. Waning influence of the farmer. M. Sullivan.

56:505-11, Sept. 1928. Vanishing farmer; why we have an agricultural problem. T. N. Carver.